LONG-THROATED FLUMES AND BROAD-CRESTED WEIRS

Long-Throated Flumes and Broad-Crested Weirs

By

M.G. BOS
International Institute for Land Reclamation and Improvement
Wageningen, The Netherlands

1985 **MARTINUS NIJHOFF/DR W. JUNK PUBLISHERS**
a member of the KLUWER ACADEMIC PUBLISHERS GROUP
DORDRECHT / BOSTON / LANCASTER

Distributors

for the United States and Canada: Kluwer Academic Publishers, 190 Old Derby Street, Hingham, MA 02043, USA
for the UK and Ireland: Kluwer Academic Publishers, MTP Press Limited, Falcon House, Queen Square, Lancaster LA1 1RN, UK
for all other countries: Kluwer Academic Publishers Group, Distribution Center, P.O. Box 322, 3300 AH Dordrecht, The Netherlands

Library of Congress Catalog Card Number: 84-25448

ISBN-13: 978-94-009-6227-9 e-ISBN-13: 978-94-009-6225-5
DOI: 10.1007/978-94-009-6225-5

Book Information

Cover photo: Portable RBC flume in a natural stream

Keywords

channels; design method; fluid flow; flumes; history; hydraulics; research; sediments; weirs

Abstract

Vital for water management are structures that can measure the flow in a wide variety of channels. Chapter 1 introduces the long-throated flume and the broad-crested weir; it explains why this family of structures can meet the boundary conditions and hydraulic demands of most measuring sites.
Chapter 2 records the history of these structures. It describes how the hydraulic theory of flumes and weirs, and their design, developed separately. The chapter concludes by reporting recent attempts to develop a generally valid theory for any long-throated flume or broad-crested weir in any channel. The remainder of the thesis explains the steps taken to develop a procedure that yields the hydraulic dimensions and rating table of the appropriate weir or flume. The major steps cover the hydraulic theory of flow through control sections of different shapes and dimensions, the theory and procedure of estimating the head loss required for modular flow, the boundary conditions of the channel, and the demands placed on the structure regarding the range and accuracy of its flow measurements.

REFERENCE: BOS, M.G., Long-throated flumes and broad-crested weirs, Martinus Nijhoff/Dr W. Junk Publishers, Dordrecht, The Netherlands, 1985

Preface

In the context of water management, structures that measure the flow rate in open channels are used for a variety of purposes:

(i) In hydrology, they measure the discharge from catchments;

(ii) In irrigation, they measure and control the distribution of water at canal bifurcations and at off-take structures;

(iii) In sanitary engineering, they measure the flow from urban areas and industries into the drainage system;

(iv) In both irrigation and drainage, they can control the upstream water at a desired level.

This thesis represents an attempt to place such flow measurements on a solid foundation by explaining the theory of water flow through 'long-throated flumes' and their hydraulically related 'broad-crested weirs'. On the basis of this theory, and on practical experience, these structures are recommended for use whenever the water surface in the channel at the measuring site can remain free.
The thesis concludes with a design procedure that will facilitate the application of these structures.

The idea to undertake research on discharge measurement structures was born upon my appointment as the first civil engineer with the International Institute for Land Reclamation and Improvement (ILRI). Because my recruiter, Ir. J.M. van Staveren, then Director of ILRI, had left the Institute before my arrival, and because I was the 'first of my kind' in an agricultural environment, I looked for contact and found support from: the late Prof. Ir. J. Nugteren, Prof. Ir. D.A. Kraijenhoff van de Leur, and Ir. R.H. Pitlo of the University of Agriculture and from Ir. J. Wijdieks, Ir. A.H. de Vries, and Ing. W. Boiten of the Delft Hydraulic Laboratory. We pooled our efforts on discharge measurement structures in the informal 'Working Group on Small Hydraulic Structures'. This thesis is founded on the pleasant and fruitful cooperation within that Working Group.

The opportunity I was given to cooperate with Dr. J.A. Replogle and A.J. Clemmens, M.Sc., of the U.S. Water Conservation Laboratory, Phoenix, on a book on long-throated flumes in open channel systems greatly expanded my knowledge of this subject. I am indebted to Dr. Ir. H. Brouwer, Director of the U.S. Water Conservation Laboratory, and to Ir. F.E. Schulze and Dr. Ir. J.A.H. Hendriks, former and present Director of ILRI, for their interest in that research program and for their efforts to ensure the required funding.

I express my gratitude first and foremost, to my promotor Prof. Ir. D.A. Kraijenhoff van de Leur, for his gift of compelling me to discuss with him every concept and detail of this thesis. If the thesis is without errors, it is because of his influence on my writing.

The thesis required the outside support and skills of various people. I am very grateful to Prof. Dr. Ing. G. Garbrecht of the Leichtweiss Institute for Water Research, Braunschweig, for reading through Chapter 2, to Dr. Ph. Th. Stol of the Institute for Land and Water Management Research (ICW) for reading and discussing Section 3.5.2, and to Dr. J.A. Replogle of the U.S. Water Conservation Laboratory for reading and discussing the entire thesis.

Special thanks are due to Ir. J.H.A. Wijbenga and Ing. J. Driegen of the Delft Hydraulics Laboratory for performing the laboratory test on the passage of sediments through weirs and flumes, and for their throughgoing discussions of this subject.

I am grateful to ILRI for providing me with the opportunity to write the thesis. I thank Prof.Dr. N.A. de Ridder for encouraging me to do so and for his positive comments when I needed them. I thank Mrs. M.F.L. Wiersma-Roche for editing the manuscript, Mr. J. van Dijk, who designed the cover and drew the figures, Mrs. G.W.C. Pleijsant-Paes and Mrs. J.B.H. van Dillen, who typed all the drafts and arranged the lay-out, to Mr. J. Ariese and Ir. M.C. van Son (ICW) who reproduced all figures photographically, and to Mr. J. van Manen and Miss E.A. Rijksen for the final production of the book. I am indebted to their great collaboration in meeting the necessary time schedule.

There are more debts to be paid. I thank my mother and late father for sending me to 'study something to do with water' at Delft after I had decided not to join the merchant marine. Of course, I am most grateful to my wife, Joke, for her resourcefulness and independence in managing our household while I was abroad. I hope that our sons Steven, Michiel, and Gijs will understand why their father was not around when he was playing with water.

Table of contents

1
The advantages of using broad-crested weirs and long-throated flumes

1.1 INTRODUCTION

Structures built for the purpose of measuring or regulating the rate of flow in open channels usually consist of a converging transition where subcritically flowing water is accelerating, a throat where it accelerates to super-critical flow, and a downstream transition where the flow velocity is reduced to an acceptable sub-critical velocity.

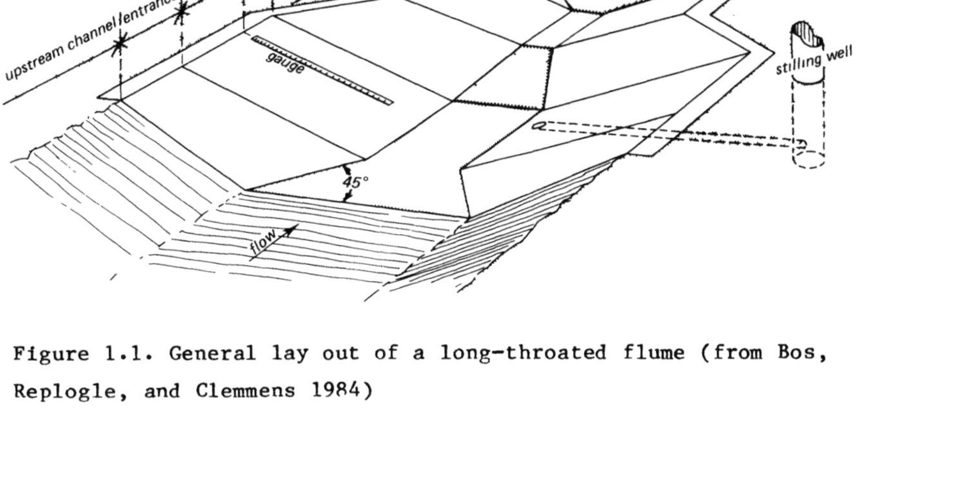

Figure 1.1. General lay out of a long-throated flume (from Bos, Replogle, and Clemmens 1984)

Upstream of the structure is an approach channel, which influences the velocity distribution of the flow approaching the structure. Downstream of the structure is a tailwater channel, which is fundamentally important to the design of the structure because the range of tailwater levels that will result from varying flow rates determines the elevation of the crest of the throat above the tailwater channel bottom.
If this tailwater is sufficiently low (Chapter 4), a diverging transition is not needed so that the structure may be truncated at the downstream end of the throat. The measuring or regulating structure can also be combined with a drop structure. If so, an energy dissipator should be added between the throat and the tailwater channel.

The difference in elevation between the water level in the approach channel, some distance upstream of the structure and the crest or invert of the horizontal throat, is known as the 'upstream sill-referenced head'. That section of the approach channel where this water surface elevation is measured is known as the 'head measurement section' or 'gauging station'.

The terms 'broad-crested weir' and 'long-throated flume' are used for two branches of the same hydraulic family of structures. Both are characterized by a throat or crest that is horizontal in the direction of flow. The difference between them is that a weir has a throat bottom which is higher than the bottom of the approach channel, whereas a flume is formed by a narrowing of the channel only. If flow is controlled by a throat that both raises the channel bottom and narrows the channel, the structure is usually called a flume (see Figure 1.2). There are, however, several structures that may be named either weir or flume.
Above the throat, the deviation from a hydrostatic pressure distribution because of centripetal acceleration may be neglected because the streamlines are practically straight and parallel. To obtain this situation, the length of the throat in the direction of flow (L) should be related to the sill-referenced energy head (H_1) as $0.10 < H_1/L < 0.50$ (Chapter 3).

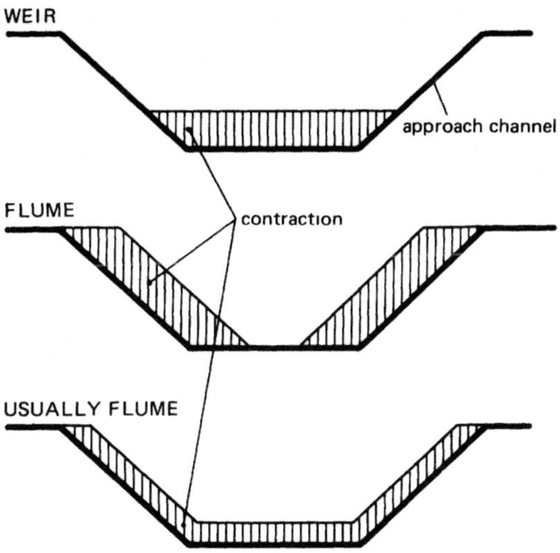

cross sections are through control at weir crest or flume throat

Figure 1.2. Distinction between a weir and a flume (from Bos, Replogle and Clemmens 1984).

1.2 ADVANTAGES

The use of broad-crested weirs or long-throated flumes is recommended for measuring flow in open channels whenever the water surface can remain free. This recommendation is made because this family of structures has the following major advantages over any other known weir or flume (Cone 1917; Inglis 1928; Jameson 1930; Bos, Replogle and Clemmens 1984):

a) Provided that critical flow occurs in the throat, a rating table can be calculated with an error of less than 5% in the listed discharge. This can be done for any combination of a prismatic throat and an arbitrarily-shaped approach channel (Chapter 3);
b) The throat, perpendicular to the direction of flow, can be shaped in such a way that the complete range of discharges can be measured accurately, and without creating an excessive backwater effect;
c) The headloss over the weir or flume required to obtain modularity - i.e. a unique relationship between the upstream sill-referenced head, h_1, and the discharge, Q - is minimal (Chapter 4);

d) This head-loss requirement can be estimated with sufficient accuracy for any of these structures placed in any arbitrary channel (Chapter 4);

e) Because of their gradually converging transitions, these structures have few problems with floating debris;

f) Field and laboratory observations have shown that the structures can be designed to pass sediment transported by channels with subcritical flow;

g) Provided that their throat is horizontal in the direction of flow, a rating table based upon post-construction dimensions can be produced, even if errors were made in constructing to the designed dimensions. Such post-construction rating also allows the throat to be reshaped, if required;

h) Under similar hydraulic and other boundary conditions, these weirs/ flumes are usually the most economical of all structures for the accurate measurement of flow.

As these advantages are being increasingly recognized, the use of long-throated flumes and broad-crested weirs is propagating. In the context of water management, they are serving a variety of purposes;

(i) In hydrology, they measure the discharge from catchments;

(ii) In irrigation, they measure and control the distribution of water at canal bifurcations and at off-take structures;

(iii) In sanitary engineering, they measure the flow from urban areas and industries into the drainage system;

(iv) In both irrigation and drainage, they can control the upstream water at a desired level.

2
History of flumes

2.1 INTRODUCTION

The world's irrigated area has increased with its population. Around 1800, only some 10 million ha were irrigated, and some 16 million ha in 1900 (James, Hanks, and Jurinak 1982). The irrigated area increased rapidly during the period 1920-40, and remained about constant during World War II and the subsequent period of decolonization. By 1955, the irrigated area was about 120 million ha (Gulhati 1955), and has now increased to about 200 million ha (FAO 1979).

Before 1900, a common method of drawing water from the conveyance system was to make an open cut in the banks of the irrigation canals. Gradually, these cuts were replaced by some form of pipe or barrel outlet. Mahbub and Gulhati (1951) give a good summary of a survey made in 1893 by the Chief Engineer of Irrigation in the Punjab.
With the increase of the irrigated area, numerous attempts were made to develop an outlet structure that provided a unique relationship between the upstream head and the discharge through the outlet. Since about 1920, the various irrigated regions have yielded various structures, amongst which were several shapes of long-throated flumes and broad-crested weirs. Many of these structures are still in use today.
With the rapid increase of the irrigated area since 1960, the need to use irrigation water more efficiently became apparent, giving new impulses to the hydraulic research on discharge measuring and regulating structures.

2.2 FLUMES WITH PIEZOMETER TAP IN THE CONVERGING TRANSITION

In the search for an 'ideal' measuring device, experiments made in 1915 in the hydraulic laboratory at Fort Collins, Colorado, led to the development of the 'Venturi flume' of Figure 2.1.
Cone (1917) selected this form and shape of device because it was 'most nearly ideal'. In his words: 'This device was accurate in its

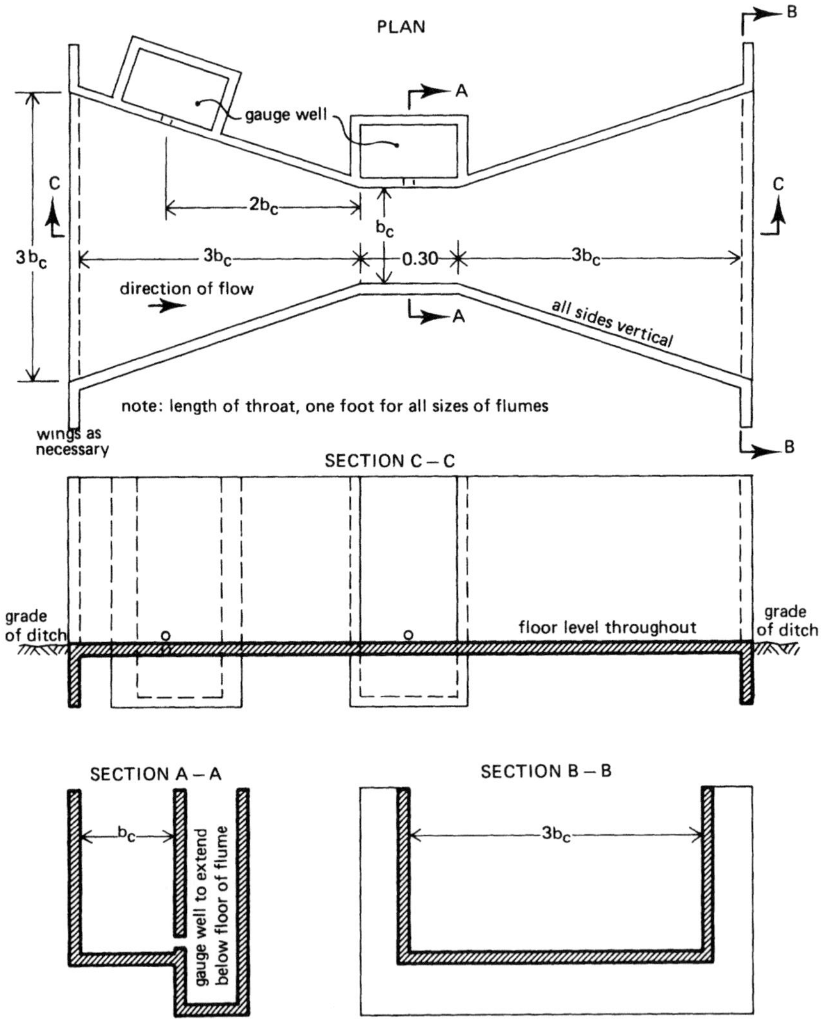

Figure 2.1. Plans for the 'Venturi flume' with rectangular control section (Cone 1917).

measurements, was free from sand, silt, or floating trash troubles, and required but little loss of head in the ditch'.

During the experiments at Fort Collins, the foundation was laid for what were to become the two typical American characteristics of a discharge measuring flume:

(i) All flumes are calibrated against a piezometric head at a prescribed location within the converging transition, the pressure tap being located at a distance of two-thirds of the transition length upstream of the leading edge of the throat;

(ii) Flumes of different capacity are obtained by changing the throat width, its length remaining constant; L = 0.305 m (1 ft) in Figure 2.1.

In his papers, Cone (1916, 1917) used the following empirical equation for modular flow through a rectangular throated flume:

$$Q = c\, b_c h_a^{\,u} \qquad (2.1)$$

where:

- c = an empirical discharge coefficient (dimensional);
- b_c = bottom width of the control section;
- h_a = sill-referenced head as measured in the upstream gauge well;
- u = empirical power ($\simeq$ 1.6).

If the head loss over the flume is limited to such an extent that the water level in the tailwater channel starts to influence the upstream head, flow is non-modular (Section 4.1). For these non-modular flows, Cone measured heads in the two gauge wells shown in Figure 2.1 and presented graphs to find the flow rate for throat widths of b_c = 0.305, 0.457, and 0.610 m.

Although Cone reported that the flume had all the qualities to become an ideal device, he also stated: 'It is not probable that the last word has been said on the design of the Venturi (long-throated) flume, for, although it has considerable promise, changes in details may prove to be necessary'.

Instead of pursuing these changes in details, Parshall (1926) further elaborated a design that had been tested and discarded by Cone (1915/16) in favour of the flume of Figure 2.1. Parshall's design had a downward chute in the throat and an upward sloping diverging transition. In 1931, this 'improved Venturi flume' was approved by the Committee on Irrigation Hydraulics of the American Society of Civil Engineers, and was named the 'Parshall measuring flume'. A comprehensive description of this flume, with dimensions and rating tables in metric units, can be found in Bos (1976).
The major disadvantages of the Parshall flume as compared with Cone's design of 1917 are;

(i) The number of plane surfaces to be constructed accurately is greater, making the flume costly to build;
(ii) In the long-throated flume, the control section is in the throat at a (variable) distance of about L/3 from the end of the throat, whereas in the Parshall flume it is at the sharp leading edge of the throat. Although a sharp edge makes a good control section, any rounding of, or damage to this sharp edge results in a systematic error in the measured discharge;
(iii) Water leaving the throat is guided towards the floor of the diverging transition, resulting in a 3 to 4 times higher head loss requirement for modular flow;
(iv) Flow at the control section is three-dimensional, for which no theoretical basis for a head-discharge equation is available.

To save on the construction cost of measuring flumes, Skogerboe et al. (1967) proposed a different modification to Cone's original design. The throat of the flume, in which the head-discharge relationship is based upon near two-dimensional flow (Chapter 3), was eliminated. The result was a flat bottomed 'Cut-throat Flume' having a converging and diverging transition only, the side walls of which intersect at a truly sharp edge. Flumes with a similar geometry were tested by Harvey (1912) in the Punjab, and later by Blau (1960) in the Democratic Republic of Germany. Both sources, however, related the flow rate to a sill-referenced head measured in the approach channel. But, any rounding of, or damage to the sharp edges of the control, any deviation in construction from the planned flume dimensions, and any changes in the flow pattern in either the approach or tailwater channel, influence the three-dimensional flow pattern in the flume. Because of the unknown sensitivity of this structure to such incidental influences, the flow pattern in the structure must be checked regularly. If necessary, the flume must be calibrated in the field or in a laboratory scale model. This calibration more than offsets the above-mentioned saving in the construction cost.

2.3 STRUCTURES WITH HEAD MEASUREMENT UPSTREAM OF THE CONVERGING TRANSITION

2.3.1 Evolution of the structure until about 1940

The broad-crested weir, with the longitudinal section of Figure 2.2, was tested by Bazin (1888). Experimental data for weirs with crest lengths between L = 0.20 and 0.40 m, and with slopes of the diverging transition ranging from 1-to-1 through 6-to-1, were published. Test results were compared on the basis of the equation of Poleni (1717), which is valid for a structure with rectangular control section:

$$Q = m \frac{2}{3} (2g)^{0.50} b_c h_1^{1.50} \qquad (2.2)$$

where:

Q = rate of flow, m^3/s;

g = acceleration due to gravity, 9.81 m/s^2;

b_c = bottom width of control section, m;

h_1 = upstream sill-referenced head, m;

m = discharge coefficient, dimensionless.

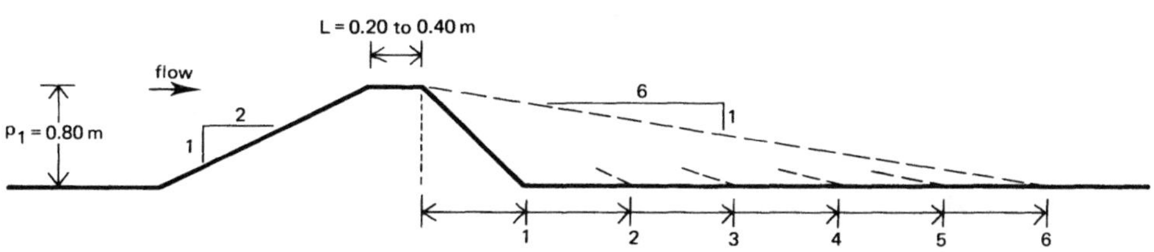

Figure 2.2. Longitudinal section over weir tested by Bazin, 1888

Equation 2.2 was derived for a sharp-crested weir under the non-realistic assumption that such a weir behaves like an orifice with a free water surface at the centre line (Rouse and Ince 1957; Bos 1976). For long-throated flumes and broad-crested weirs, this formula nowadays has been abandoned in favour of Eq. 2.3, which is based on the sounder assumption of critical flow in open channels.

The experiments of Bazin were widely discussed in German, English, and American literature (Gravelius 1900; Rafter 1900; Horton 1907) and can be regarded as an important contribution to the search for the 'ideal' irrigation water measuring weir.

Although Bélanger (1849) and Bazin (1888) were fully aware of the existence of a unique relationship between the upstream sill-referenced head and the discharge over a weir with critical flow at the control section, their hydraulic theory was not used by irrigation engineers in their search for a discharge measuring flume. A possible explanation for this is that hydraulic engineers regarded weirs and flumes as structures with different hydraulic behaviour.

In Colorado, Cone (1917) laid the foundation for the U.S. practice of designing a style of flumes with a piezometer tap (Section 2.2).

A second style of 'Venturi flume' was originated by the irrigation engineers, Harvey and Stoddard, both of the Punjab Irrigation Department (1912). Lindley (1931) described their flume as follows: 'By introducing a smooth hump on the bed, or smooth contractions of the sides of a regular channel, a dip in the water surface is produced equivalent to the head converted into increased velocity. Knowing the areas of the upstream and throat sections, and the amount of this dip, the discharge can be calculated' (Section 3.1).

The Venturi flume requires that both the upstream sill-referenced head and the lowest head in the throat be measured. Because the horizontal location where the lowest head (water surface dip) occurs in the throat is variable with head and flow rate, a long-throated flume was developed in which critical flow occurred. Thus a unique relationship between the upstream sill-referenced head and the 'modular' rate of flow was established. A downstream 'standing wave' guaranteed the presence of critical flow in the throat. Inglis (1928) reported on Crump's experiments on these 'standing wave flumes'. The head-discharge equation of the long-throated (standing wave) flume, derived by Bélanger (1848) and confirmed by Crump, reads:

$$Q = C \frac{2}{3} (\frac{2}{3} g)^{0.50} b_c h_1^{1.50} \qquad (2.3)$$

This equation is almost identical to Eq. 3.18, which today is generally accepted.

The hydraulic theory of long-throated flumes and broad-crested weirs was further advanced by Jameson (1925, 1930), who derived the general equations for ideal flow (Eqs. 3.8 and 3.12).

Parallel to this hydraulic development, field experience (Lindley 1925) and laboratory experiments (Fane 1927) led to the classic flume as described by Inglis (1928). This long-throated flume, illustrated in Figure 2.3, became very popular for irrigation water measurement. Mahbub and Gulhati (1951) report that, in 1944, about 12,500 of these open flumes were being used in canals in the Punjab alone.

The first systematic research on the minimum head loss required for critical flow in long-throated flumes with rectangular control section and with diverging transitions of different flare angles and lengths was reported by Fane (1927). This minimum required head loss for flow to remain 'modular' was expressed in the maximum allowable submergence ratio H_2/H_1 (Section 4.1). Based on Fane's tests, and earlier experiments by Crump, Inglis (1928) published the data of Table 2.1.

Fane (1927) stated that, to obtain these modular limits, the energy loss due to turbulence in the zone of deceleration and to friction should be minimized. He noted that very gradual, but long, diverging transitions effectively suppress turbulence but have relatively high energy losses due to friction and thus may have a lower modular limit than those listed in Table 2.1 (see also Bos and Reinink 1981 and Chapter 4). To reduce the construction cost of the upstream converging transition, Fane (1927) gave the following design rules: 'It is better to have the throat as long as possible ($L = 2H_{1max}$), considering expense, and to have short upstream wing walls, say 0.10 m radius.'

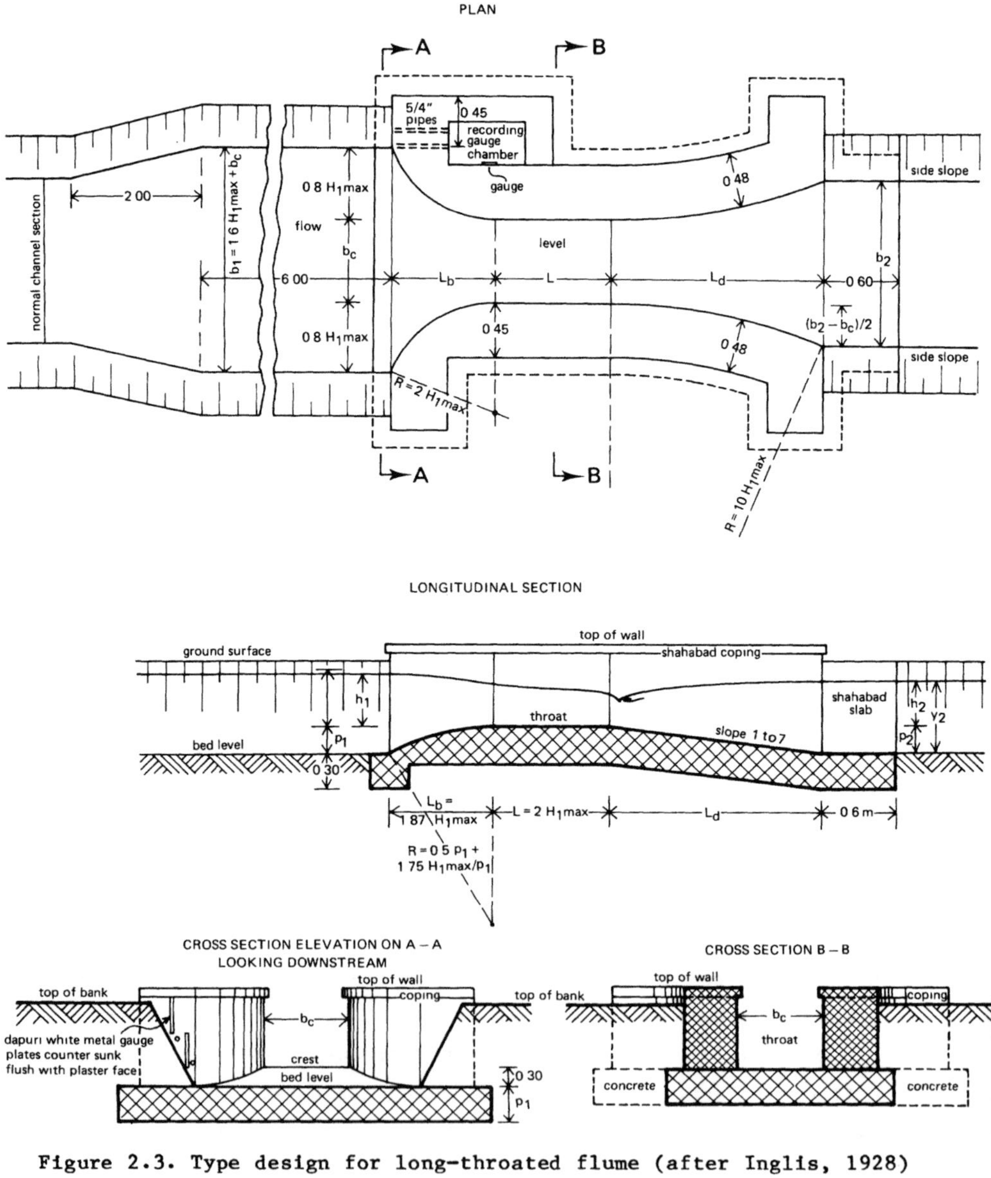

Figure 2.3. Type design for long-throated flume (after Inglis, 1928)

This radius equals about 0.15 h_{1max}. (H_{1max} denotes the maximum anticipated value of the upstream sill-referenced energy head.)

Table 2.1. Modular limits for a long-throated flume (Inglis 1928)

Bottom slope of diverging transition	Modular limit (per cent)
5 to 1	80
10 to 1	85
15 to 1	88
20 to 1	94

To measure and regulate irrigation water on Java, Romijn (1932) tested a broad-crested weir with the longitudinal profile shown in Figure 2.4.

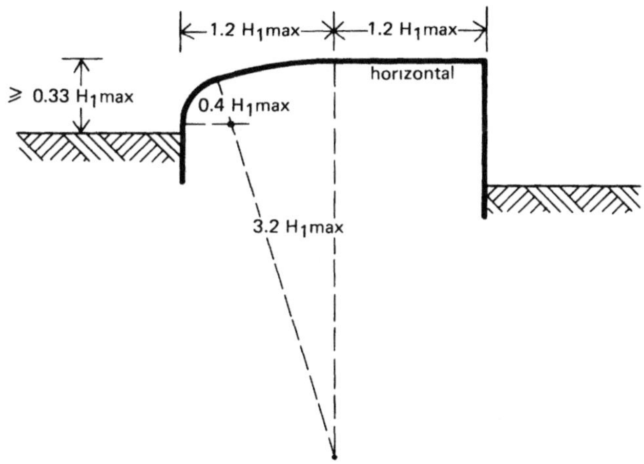

Figure 2.4. Longitudinal profile over a Romijn weir (1932)

Although Romijn designed a compact weir profile, the above recommendations of Fane yield a shorter structure. Romijn's weir crest was mounted on top of a vertically movable gate so that the crest elevation could be changed with respect to the upstream water level. The resulting structure was an excellent discharge measuring and regulating device which is still widely used today.

Later, in 1936, Palmer and Bowlus developed some trapezoidal flumes for sewer flow measurement. These flumes, which still bear their name, were further refined by Arredi (1936) and by Wells and Gotaas (1958). Although earlier researchers have used curved converging transitions, Palmer and Bowlus used plane transitions, which have the merit of simple construction. Wells and Gotaas showed that the difference in performance between structures having either a curved or an 1-to-3 plane converging transition is negligible. In 1981, Bos and Reinink showed that a 1-to-2 transition is sufficient to avoid flow separation at the leading edge of the throat. This, in fact, is the sole function of this transition (Section 3.5.3). It is interesting to note that the latter development yielded a longitudinal section over the structure similar to that of Bazin (Figures 2.2 and 2.5).

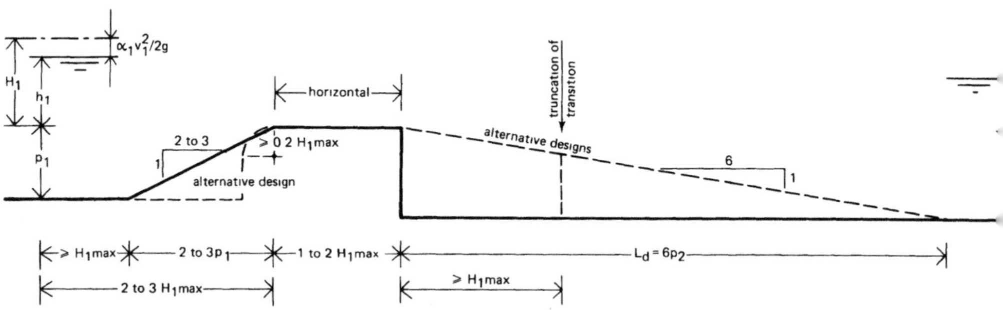

Figure 2.5. Longitudinal profile of a broad-crested weir or a long-throated flume (Bos, Replogle, and Clemmens 1984)

2.3.2 Discharge rating based on boundary layer development

In evidence of the growing importance being attached to water control in newly designed irrigation projects, further systematic experimental work on broad-crested weirs and long-throated flumes was conducted again after about 1955 (Wells and Gotaas 1956; Robinson and Chamberlain 1960; Hall 1962, 1967).

Ackers and Harrison (1963) applied the theory of boundary layer development to flow in the flume throat. They produced a general design method for flumes with a trapezoidal control section, including the two limiting cases, namely the rectangular and triangular control. Their paper laid the foundation for the head-discharge equation:

$$Q = C_d C_v C_s \frac{2}{3} \left(\frac{2}{3}g\right)^{0.50} b_c h_1^{1.50} \tag{2.4}$$

where:

C_d = discharge coefficient (Section 3.5);

C_v = approach velocity coefficient (Section 3.6);

C_s = a shape coefficient which corrects for the control section not being rectangular;

b_c = bottom width of control section;

h_1 = sill-referenced head.

This general design method introduced the boundary layer displacement depth, a shape coefficient, and an iteration procedure to calculate a rating table with the exclusive use of structure dimensions. This method was accepted by the British Standards Institution (1974) and by the International Standards Organization (1980) as a cornerstone for their standards on long-throated flumes.

The above method of producing a rating table assumes that the velocity distributions at the gauging station and in the throat (control) section, reduced by the displacement depths, are uniform. A non-uniform velocity distribution due to streamline curvature, however, can influence the flume rating by approximately as much as the introduced boundary layer displacement depth. Replogle (1975) reported the Ackers and Harrison method to be inconsistent in accurately predicting the laboratory calibration of a variety of carefully tested flumes. This prompted the development of a computer model for making a rating table that uses both the boundary layer theory and a prediction of the velocity distribution. Replogle (1978) states that this model for broad-crested weirs and long-throated flumes, having any arbitrary shape of the control section and being placed in any channel, produces the calibration with an error in the listed discharge of 2% or less.

2.3.3 Discharge rating based on H_1/L ratio

As an expansion to the hydraulic theory presented by Jameson (1925) and to the method of allowing for the velocity of approach, C_v, as used by the British Standards Institution (1969), a series of head-discharge equations were developed (Section 3.3).
As shown in Figure 2.6, all equations use a discharge coefficient, C_d, and either an approach velocity coefficient, C_v, or a table to determine the rate of the flow.

In the practice of irrigation, and of water management in general, the flows to be measured are within a certain range. Bos (1976, 1977) and Clemmens, Bos, and Replogle (1984) found that within the range $0.1 < H_1/L < 1.0$, one relationship between the discharge coefficient, C_d, and the dimensionless ratio, H_1/L, is sufficiently accurate for all long-throated flumes and broad-crested weirs (Section 3.5.2)

Bos (1977) also developed a generally valid procedure which gives values of the approach velocity coefficient for all combinations of control shape and approach channel (Section 3.6).

With the equations as listed in Figure 3.12 and the above values for C_d and C_v, a rating table for the related control sections can be produced.

2.3.4 Estimate of the modular limit

In flat irrigated areas, the measurement and controlled distribution of flow over a number of successive canal bifurcations requires a considerable part of the available head in the canal system. An accurate estimate of this part of the available head results in a good knowledge of the available hydraulic gradient along the reaches of the canal system. For an individual structure, a correct estimate of the required head loss, as a function of the hydraulic design of that structure, is essential for determining the crest or throat elevation with respect to the tailwater level (Chapter 4). This is true for any measuring structure in a stream with a flat hydraulic gradient.

As mentioned when Table 2.1 was being introduced, the first systematic research on the head loss requirement for modular flow was reported by Fane (1927). Subsequently, other researchers (Engel 1934; Wells and Gotaas 1956; Blau 1960; Landbouwhogeschool Wageningen 1964 to 1971; Harrison 1968; and Smith and Liang 1969) published data on the modular limit of individual structures. No general system of estimating the modular limit became available, and the guidelines published by the British Standards Institution (1974) and by the International Standards Organization (1980, 1982) are in the same tabular form as those published by Inglis in 1928 (Table 2.1.)

In 1976, Bos published a method of estimating the modular limit of a structure having any arbitrary shape and being placed in any channel. This 16-step method was laboratory tested and checked against published data (Bos and Reinink 1981). A recent study on the performance of a large weir (16.45 m wide) and recent laboratory tests showed that the method is sufficiently accurate and reliable for the hydraulic design of broad-crested weirs and long-throated flumes (Replogle et al. 1983; Dodge 1982; Boiten 1983). A detailed description of this method is given in Chapter 4.

3
The head-discharge relationship

3.1 BASIC EQUATIONS

Figure 3.1 shows a short zone of acceleration bounded by the section of the gauging station and the control section. Both sections are perpendicular to straight and parallel streamlines.

In both cross-sections, the effect of streamline curvature on the piezometric level is negligible so that this level coincides with the water surface levels. These water levels determine the areas A_1 and A_c.

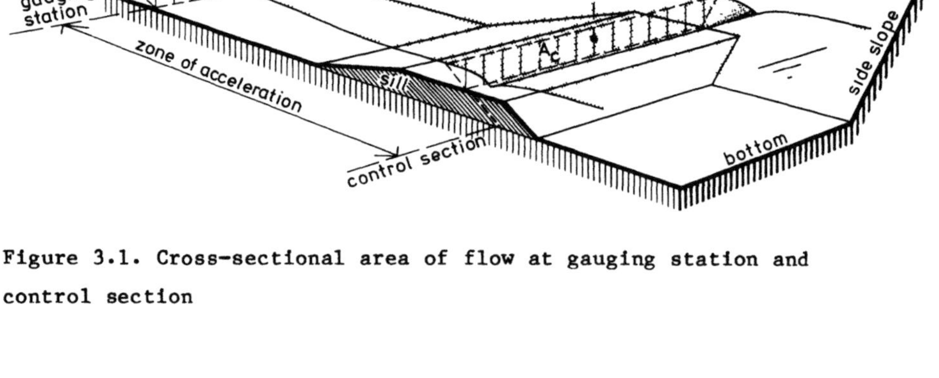

Figure 3.1. Cross-sectional area of flow at gauging station and control section

We may therefore conclude that if the shapes of the gauging section and the control section are known, and the two corresponding water levels are measured, the two unknown average velocities, v_1 and v_c, can be determined from the following equations:

Conservation of mass, if density changes are neglected:

$$Q = v_1 A_1 = v_c A_c \tag{3.1}$$

Conservation of energy, if energy losses in the zone of acceleration are neglected:

$$\alpha_1 v_1^2/2g + (P/\rho g + Z)_1 = \alpha_c v_c^2/2g + (P/\rho g + Z)_c \tag{3.2}$$

Flow measuring structures as described above, which require that two water levels be determined and converted into areas of cross-section, are named Venturi flumes. The measurement and conversion of two heads, however, is time-consuming and expensive, and should be avoided if possible. It will be shown that the measurement of one water level in the gauging section is sufficient to determine the rate of flow, provided that the flow in the control section is critical. To explain this critical flow condition, the concept of specific energy will first be defined.

3.2 SPECIFIC ENERGY

The concept of specific energy was first introduced by Bakhmeteff in 1912, and was defined as the average energy per unit of weight of water at a channel section, with the channel bottom serving as reference level. Since the piezometric level coincides with the water surface level, the piezometric head with respect to the channel bottom is:

$$P/\rho g + Z = y, \text{ the water depth} \tag{3.3}$$

So that the specific energy head is:

$$H_o = y + \alpha v^2/2g \tag{3.4}$$

Substitution of $v = Q/A$ into Eq. 3.4 yields:

$$H_o = y + \alpha\, Q^2/(2g\, A^2) \tag{3.5}$$

where A, the cross-sectional area of flow, can also be expressed as a function of y, so that for a given channel cross-section and a constant discharge, the specific energy head is a function of the water depth only. Plotting this water depth, y, against the specific energy, H_o, gives the specific energy curve as shown in Figure 3.2.

The curve shows that, for a given discharge, Q_2, and specific energy, H_o, there are two 'alternate depths' of flow: one for supercritical flow and one for subcritical flow. At point C, the specific energy is at its minimum for the given discharge, and the two alternate depths coincide. This depth of flow is known as 'critical depth', y_c. The relationship between this minimum specific energy head, H_c, and the critical depth is found by differentiating Eq. 3.5 to y, while Q remains constant.

$$dH_o/dy = 1 - (\alpha Q^2/gA^3)(dA/dy)$$
$$= 1 - (\alpha v^2/gA)(dA/dy)$$

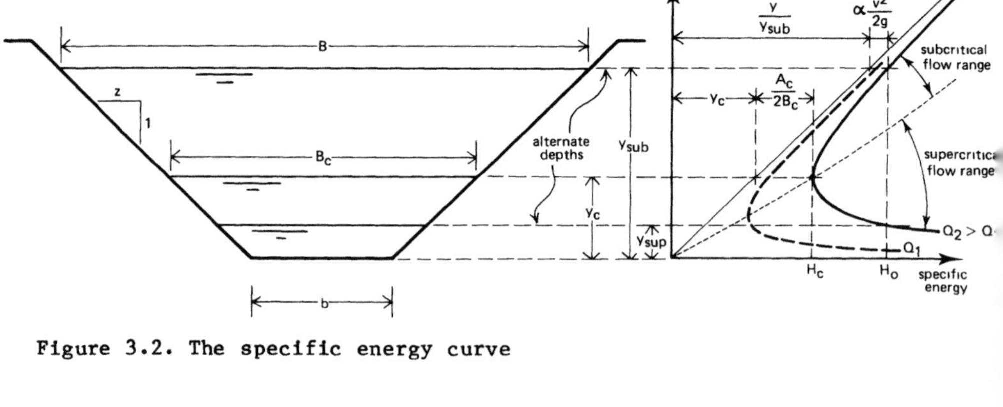

Figure 3.2. The specific energy curve

Since dA = Bdy, this equation becomes

$$dH_o/dy = 1 - \alpha v^2 B/gA \qquad (3.7)$$

If water flows at critical depth, the specific energy head is at its minimum. Thus $dH_o/dy = 0$ and Eq. 3.7 becomes (Jameson 1925):

$$\alpha_c v_c^2/2g = A_c/2B_c \qquad (3.8)$$

The critical velocity, v_c, is the maximum speed for disturbances to travel with respect to the flowing water. This implies that supercritical flow is only influenced by upstream conditions. Hence, the (supercritical) rate of flow is independent of the tailwater level; the flow over the structure is then called 'modular'. The following sections will show that Eqs. 3.1, 3.2, and 3.8 allow the development of a head-discharge relationship for modular flow in any broad-crested weir or long-throated flume.

3.3 HEAD-DISCHARGE EQUATIONS

The head-discharge equations in this section are needed for two purposes:

(i) for structure design;

(ii) for rating table or rating curve production.

When the structure is being designed, the values of h_{1min}, p_1, and h_{1max} must be matched to the anticipated flow rates Q_{min} and Q_{max}. This matching process usually requires the calculation of h_{1min} and h_{1max} for one or two control shapes and a variety of dimensions (Chapter 5).

Once the shape and dimensions of the control section have been selected, the head-discharge equations are used again to produce a h_1-Q rating of the designed structure. Often, this rating is made after the weir or flume has been constructed, so that the actual dimensions of the control section can be used. If a programmable calculator is available, the iteration method described in Section 3.7 can be used to produce the rating table or curve.

3.3.1 General equation for ideal flow

A broad-crested weir, or the related long-throated flume, is an overflow structure with a horizontal crest in the direction of flow. In the control section, the streamlines are practically straight and parallel so that the effect of centripetal acceleration on the piezometric level can be neglected and the pressure distribution is hydrostatic.
To obtain this situation, the crest length, L, should be related to the sill- or crest-referenced energy head, H_1, as:

$$H_1/L < 0.50 \tag{3.9}$$

so that curvature of streamlines does not significantly disturb the hydrostatic pressure distribution in the control section (see also Sections 3.5.2 and 4.3.3).

If the measuring structure is so designed that the converging transition leads the water into the throat (or above the crest), without significant flow separation and subsequent turbulence losses, there is a smooth flow pattern upstream of the control section. Neglecting friction losses in this zone of acceleration we thus may write (Figures 3.1 and 3.3):

$$H_1 = h_1 + \alpha_1 v_1^2/2g = H_c = y_c + \alpha_c v_c^2/2g \tag{3.10}$$

or:

$$\alpha_c v_c^2 = 2g(H_1 - y_c) \tag{3.11}$$

where H_1 equals the sill-referenced energy head as shown in Figure 3.3.

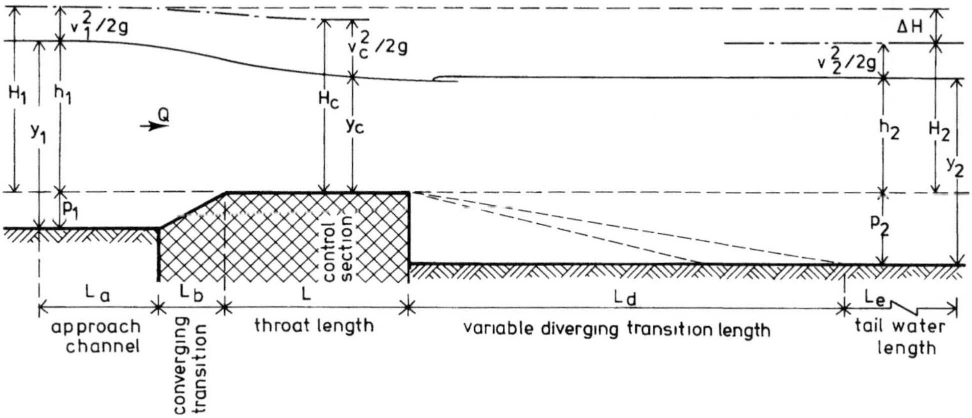

Figure 3.3. Illustration of terminology

Combining $Q = v_c A_c$ and Eq. 3.11 and assuming a practically uniform velocity distribution, $\alpha_c = 1.0$, gives:

$$Q_i = A_c \left[2g(H_1 - y_c)\right]^{0.50} \qquad (3.12)$$

In this equation for the ideal rate of flow, Q_i, we can use Eq. 3.8 to calculate:

$$y_c = H_1 - A_c/2B_c \qquad (3.13)$$

The head-discharge equation of a structure, having any well-defined control section and being placed in any channel, can be derived from Eqs. 3.12 and 3.13, as will be done in the following sections.

3.3.2 Rectangular control section

For a rectangular control section in which the flow is critical, we may write $A_c = b_c y_c$ and $B_c = b_c$. Hence:

$$y_c = H_1 - \frac{1}{2} y_c \qquad (3.14)$$

or:

$$y_c = \frac{2}{3} H_1 \qquad (3.15)$$

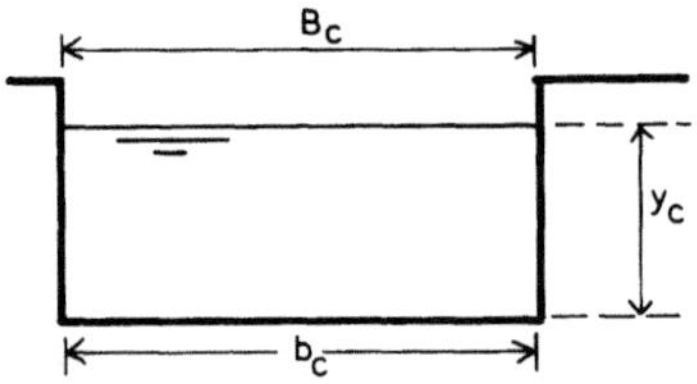

Figure 3.4. Dimensions of a rectangular control section

Substituting of this relationship and $A_c = b_c y_c$ into Eq. 3.12 and rewriting gives:

$$Q_i = \frac{2}{3} \left(\frac{2}{3}g\right)^{0.50} b_c H_1^{1.50} \tag{3.16}$$

This equation for ideal flow over a broad-crested weir was first determined by Bélanger in 1849. It is based upon a number of idealized assumptions such as: absence of energy losses between the gauging and control sections; uniform velocity distribution in both sections; and straight parallel streamlines at the gauging and control sections. In reality these assumptions are not entirely correct, but deviations from them can be compensated for by the introduction of a discharge coefficient, C_d; $Q = C_d Q_i$.

Equation 3.16 then reads:

$$Q = C_d \frac{2}{3} \left(\frac{2}{3}g\right)^{0.50} b_c H_1^{1.50} \tag{3.17}$$

In a field installation, it is impracticable to measure the energy head, H_1, directly. Common practice therefore is to relate the discharge to the upstream sill-referenced head, h_1. To correct for neglecting the velocity head, $\alpha_1 v_1^2/2g$, an approach velocity coefficient, C_v, is introduced into Eq. 3.17. Hence:

$$Q = C_d C_v \frac{2}{3} \left(\frac{2}{3}g\right)^{0.50} b_c h_1^{1.50} \tag{3.18}$$

which is the final equation for a rectangular control section. For further details on the C_d and C_v values, reference is made to Sections 3.5 and 3.6 respectively.

3.3.3 Triangular control section

For a triangular control section (Figure 3.5), we may write $A_c = y_c^2 \tan \frac{\theta}{2}$ and $B_c = 2\, y_c \tan \frac{\theta}{2}$. Substitution of these values into Eq. 3.13 gives:

$$y_c = \frac{4}{5} H_1 \qquad (3.19)$$

Substitution of this y_c value and of $A_c = y_c^2 \tan \frac{\theta}{2}$ into Eq. 3.12 gives (Jameson 1925):

$$Q = \frac{16}{25} \left(\frac{2}{5} g\right)^{0.50} \tan \frac{\theta}{2} H_1^{2.50} \qquad (3.20)$$

For the reasons explained in Section 3.3.2, the introduction of C_d and C_v completes this equation to:

$$Q = C_d C_v \frac{16}{25} \left(\frac{2}{5} g\right)^{0.50} \tan \frac{\theta}{2} h_1^{2.50} \qquad (3.21)$$

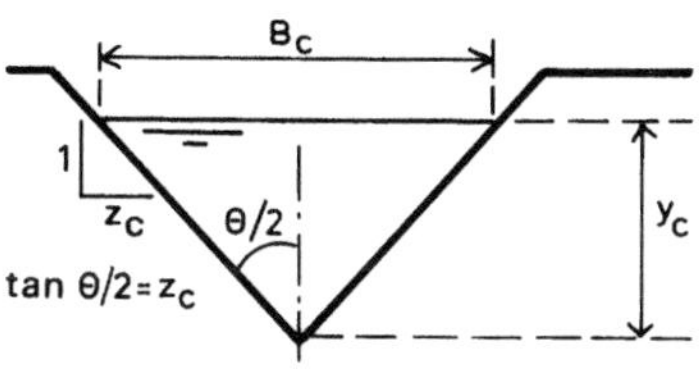

Figure 3.5. Dimensions of a triangular control section

3.3.4 Truncated triangular control section

For structures with a truncated triangular control section (Figure 3.6), two head-discharge equations should be used: one for the conditions where flow is confined within the triangular part of the control section, and the other, at higher stages, where the presence of the vertical side walls has to be taken into account. The first equation is analogous to Eq. 3.21, being:

$$Q = C_d C_v \frac{16}{25} \left(\frac{2}{5} g\right)^{0.5} \tan \frac{\theta}{2} h_1^{2.50} \qquad (3.22)$$

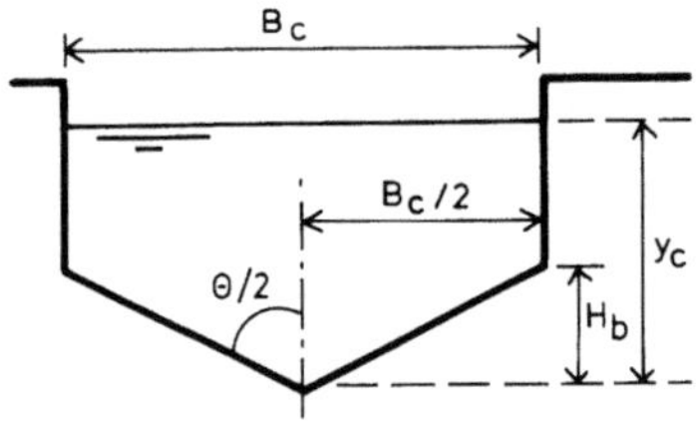

Figure 3.6. Dimensions of a truncated triangular ccntrol section

This equation is valid if $y_c < H_b$. Substitution of this limitation into Eq. 3.19 gives the limit of application of Eq. 3.22 for this control shape as (Bos 1976):

$$H_1 < 1.25\ H_b \tag{3.23}$$

A head-discharge equation for $H_1 > 1.25\ H_b$ or $y_c > H_b$ was derived by Bos in 1972 as follows:

$$A_c = B_c y_c - \frac{1}{2} B_c H_b = B_c\ (y_c - \frac{1}{2} H_b) \tag{3.24}$$

(Note that $y_c - \frac{1}{2} H_b$ is the average water depth in the control section.)

Substitution of A_c and B_c into Eq. 3.13 gives:

$$y_c = \frac{2}{3} H_1 + \frac{1}{6} H_b \text{ and } H_1 - y_c = \frac{1}{3} (H_1 - \frac{1}{2} H_b) \tag{3.25}$$

so that Eq. 3.24 can also be written as:

$$A_c = \frac{2}{3} B_c H_1 - \frac{1}{3} B_c H_b = \frac{2}{3} B_c\ (H_1 - \frac{1}{2} H_b) \tag{3.26}$$

Substituting the Eqs. 3.25 and 3.26 into Eq. 3.12 and adding the C_d and C_v values gives

$$Q = C_d C_v \frac{2}{3} (\frac{2}{3} g)^{0.50} B_c (h_1 - \frac{1}{2} H_b)^{1.50} \tag{3.27}$$

A comparison between this equation and Eq. 3.18 shows that the flow through the triangular part of the control section is accounted for by subtracting $\frac{1}{2} H_b$ from the sill-referenced head. Flow through a truncated triangular control thus equals the flow through a rectangular control of similar width and with a sill reference location at $0.5\ H_b$ above the invert of the triangle. This may be called the equivalent rectangular control section, and $h_1 - 0.5\ H_b$ is the upstream head with respect to the equivalent horizontal sill in the control section. (For other complex control sections, see Section 3.4.)

3.3.5 Trapezoidal control section

For a trapezoidal control section (Figure 3.7) with $A_c = b_c y_c + z_c y_c^2$ and $B_c = b_c + 2z_c y_c$, we may write Eq. 3.8 as for $\alpha_c = 1$:

$$\frac{v_c^2}{2g} = \frac{b_c y_c + z_c y_c^2}{2b_c + 4z_c y_c} \tag{3.28}$$

Since for ideal fluid flow $H_1 = H_c = v_c^2/2g + y_c$, we may write the total sill-referenced energy head as a function of the dimensions of the control section as

$$H_1 = \frac{3b_c y_c + 5z_c y_c^2}{2b_c + 4z_c y_c} \tag{3.29}$$

or:

$$2\frac{b_c}{H_1} + 4z_c\frac{y_c}{H_1} = 3\frac{b_c}{H_1}\frac{y_c}{H_1} + 5z_c\left(\frac{y_c}{H_1}\right)^2 \tag{3.30}$$

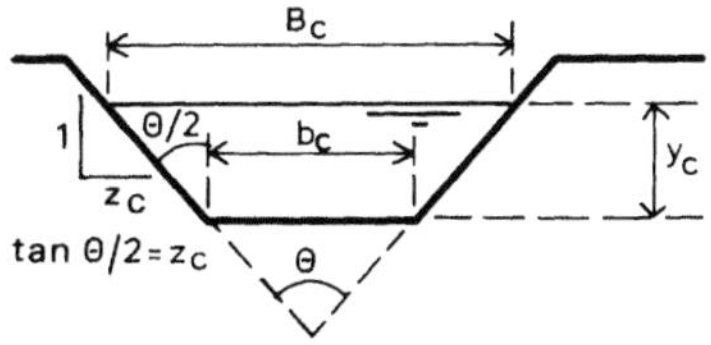

Figure 3.7. Dimensions of a trapezoidal control section

From this equation, it appears that the critical depth in the control section is a function of the energy head, H_1, of the bottom width, b_c, and of the side slope ratio, z_c, of the control section. It also appears that, if z_c is known, the ratio y_c/H_1 is a function of H_1/b_c. Values of y_c/H_1 as a function of z_c and the ratio H_1/b_c are shown in Table 3.1.

Substitution of $A_c = b_c y_c + z_c y_c^2$ into Eq. 3.12 and the introduction of a discharge coefficient give a head-discharge equation (Bos 1976):

$$Q = C_d[b_c y_c + z_c y_c^2][2g(H_1 - y_c)]^{0.50} \tag{3.31}$$

For given values of H_1, b_c, and z_c, a value for the ratio y_c/H_1 can be derived from Table 3.1, so that y_c is known and the discharge can be calculated from Eq. 3.31. By varying H_1, the Q-H_1 relationship is obtained. This Q-H_1 relationship can be converted into Q-h_1 as follows:

1. Estimate A_1 by assuming that $h_1 \simeq H_1$;
2. Calculate $h_1 = H_1 - \alpha_1 Q^2/2gA_1^2$;
3. Use this new h_1 value to calculate A_1;
4. Repeat Step 2 to find a sufficiently accurate h_1 value;
5. Plot Q-h_1 curve or make Q-h_1 rating table.

Table 3.1. Values of the ratio y_c/H_1 as a function of z_c and H_1/b_c for trapezoidal control sections

	Side slopes of channel, ratio of horizontal to vertical (z_c)									
H_1/b_c	Vertical	0.25:1	0.50:1	0.75:1	1:1	1.5:1	2:1	2.5:1	3:1	4:1
.00	.667	.667	.667	.667	.667	.667	.667	.667	.667	.667
.01	.667	.667	.667	.668	.668	.669	.670	.670	.671	.672
.02	.667	.667	.668	.669	.670	.671	.672	.674	.675	.678
.03	.667	.668	.669	.670	.671	.673	.675	.677	.679	.683
.04	.667	.668	.670	.671	.672	.675	.677	.680	.683	.687
.05	.667	.668	.670	.672	.674	.677	.680	.683	.686	.692
.06	.667	.669	.671	.673	.675	.679	.683	.686	.690	.696
.07	.667	.669	.672	.674	.676	.681	.685	.689	.693	.699
.08	.667	.670	.672	.675	.678	.683	.687	.692	.696	.703
.09	.667	.670	.673	.676	.679	.684	.690	.695	.698	.706
.10	.667	.670	.674	.677	.680	.686	.692	.697	.701	.709
.12	.667	.671	.675	.679	.684	.690	.692	.701	.706	.715
.14	.667	.672	.676	.681	.686	.693	.699	.705	.711	.720
.16	.667	.672	.678	.683	.678	.696	.703	.709	.715	.725
.18	.667	.673	.679	.684	.690	.698	.706	.713	.719	.729
.20	.667	.674	.680	.686	.692	.701	.709	.717	.723	.733
.22	.667	.674	.681	.688	.694	.704	.712	.720	.726	.736
.24	.667	.675	.683	.689	.696	.706	.715	.723	.729	.739
.26	.667	.676	.684	.691	.698	.709	.718	.725	.732	.742
.28	.667	.676	.685	.693	.699	.711	.720	.728	.734	.744
.30	.667	.677	.686	.694	.701	.713	.723	.730	.737	.747
.32	.667	.678	.687	.696	.703	.715	.725	.733	.739	.749
.34	.667	.678	.689	.697	.705	.717	.727	.735	.741	.751
.36	.667	.679	.690	.699	.706	.719	.729	.737	.743	.752
.38	.667	.680	.691	.700	.708	.721	.731	.738	.745	.754
.40	.667	.680	.692	.701	.709	.723	.733	.740	.747	.756
.42	.667	.681	.693	.703	.711	.725	.734	.742	.748	.757
.44	.667	.681	.694	.704	.712	.727	.736	.744	.750	.759
.46	.667	.682	.695	.705	.714	.728	.737	.745	.751	.760
.48	.667	.683	.696	.706	.715	.729	.739	.747	.752	.761
.5	.667	.683	.697	.708	.717	.730	.740	.748	.754	.762
.6	.667	.686	.701	.713	.723	.737	.747	.754	.759	.767
.7	.667	.688	.706	.718	.728	.742	.752	.758	.764	.771
.8	.667	.692	.709	.723	.732	.746	.756	.762	.767	.774
.9	.667	.694	.713	.727	.737	.750	.759	.766	.770	.776
1.0	.667	.697	.717	.730	.740	.754	.762	.768	.773	.778
1.2	.667	.701	.723	.737	.747	.759	.767	.772	.776	.782
1.4	.667	.706	.729	.742	.752	.764	.771	.776	.779	.784
1.6	.667	.709	.733	.747	.756	.767	.774	.778	.781	.786
1.8	.667	.713	.737	.750	.759	.770	.776	.781	.783	.787
2	.667	.717	.740	.754	.762	.773	.778	.782	.785	.788
3	.667	.730	.753	.766	.773	.781	.785	.787	.790	.792
4	.667	.740	.762	.773	.778	.785	.788	.790	.792	.794
5	.667	.748	.768	.777	.782	.788	.791	.792	.794	.795
10	.667	.768	.782	.788	.791	.794	.795	.796	.797	.798
∞		.800	.800	.800	.800	.800	.800	.800	.800	.800

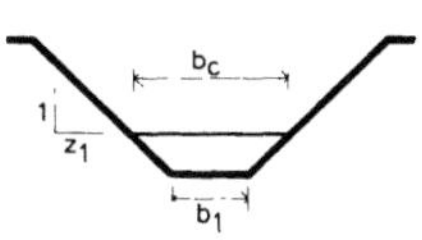

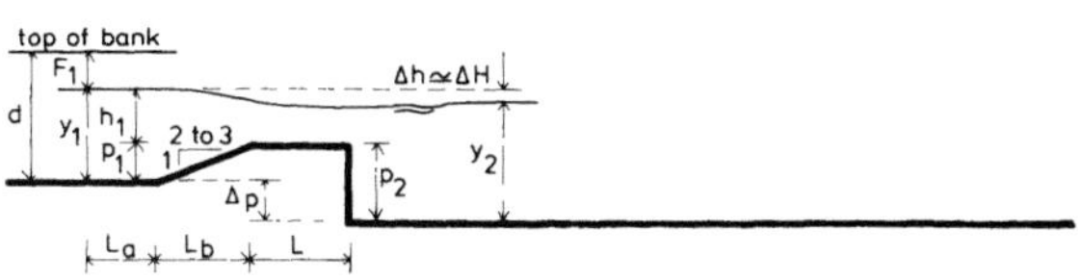

3.3.6 Parabolic control section

The relatively uncommon parabolic control section (Figure 3.8) can be built from a section of a prefabricated irrigation canal. If f_c is the focal distance of the parabola, we can write $A_c = \frac{2}{3} B_c y_c$ and $B_c = 2\sqrt{2f_c y_c}$.

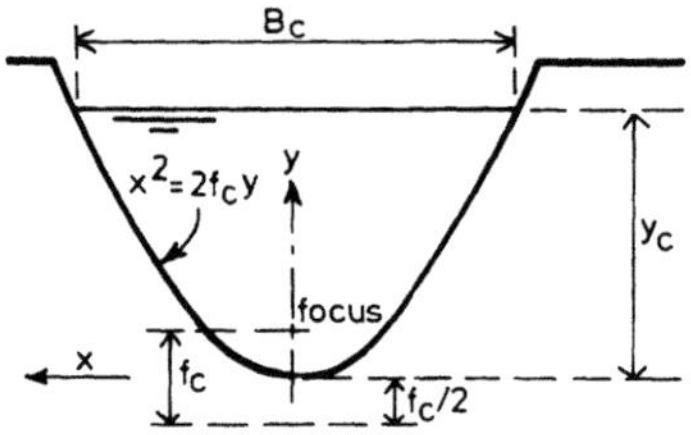

Figure 3.8. Dimensions of a parabolic control section

Substitution of A_c into Eq. 3.13 yields:

$$y_c = \frac{3}{4} H_1 \qquad (3.32)$$

Substitution into Eq. 3.12 of this y_c value, $A_c = \frac{2}{3} B_c y_c$, and $B_c = 2\sqrt{2f_c y_c}$, and the introduction of C_d and C_v gives (Jameson 1925):

$$Q = C_d C_v \left(\frac{3}{4} g f_c\right)^{0.50} h_1^{2.0} \qquad (3.33)$$

3.3.7 Circular control section

For a structure with a circular control section (Figure 3.9) we may write:

$$A_c = \frac{1}{8} d_c^2 (\Theta - \sin \Theta) \qquad (3.34)$$

$$B_c = d_c \sin \frac{\Theta}{2} \qquad (3.35)$$

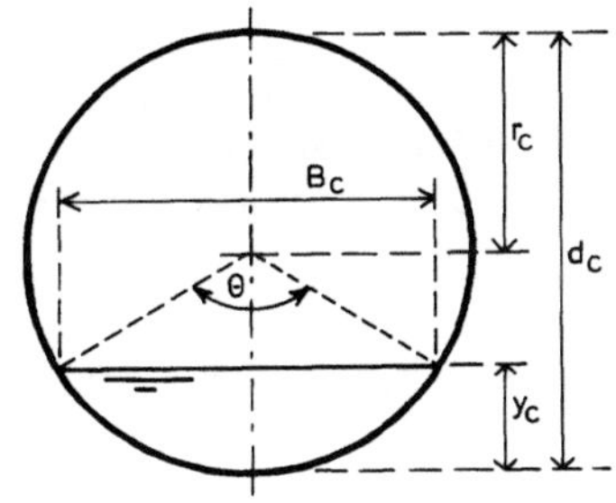

Figure 3.9. Dimensions of circular control section

and:

$$y_c = \frac{1}{2}d_c\ (1 - \cos\frac{\theta}{2}) = d_c \sin^2\frac{\theta}{4} \tag{3.36}$$

Substitution of A_c and B_c into Eq. 3.8 gives

$$\frac{v_c^2}{2g} = \frac{d_c}{16}\ \frac{\theta - \sin\theta}{\sin\frac{\theta}{2}} \tag{3.37}$$

and because $H_1 = H_c = y_c + v_c^2/2g$, we may write a dimensionless form of the sill-referenced energy head as:

$$\frac{H_1}{d_c} = \frac{y_c}{d_c} + \frac{v_c^2}{2gd_c} = \sin^2\frac{\theta}{4} + \frac{\theta - \sin\theta}{16\sin\frac{\theta}{2}} \tag{3.38}$$

For each value of the dimensionless ratio $y_c/d_c = \sin^2\frac{\theta}{4}$, matching values of the ratios A_c/d_c^2 and H_1/d_c can be calculated with the above information. These values, and the additional values of the dimensionless ratios, $v_c^2/2gd_c$ and y_c/H_1, are presented in Table 3.2.

For a circular control section, we can use the general head-discharge equation given earlier (Eq. 3.12) and write

$$Q = C_d A_c\ \{2g(H_1 - y_c)\}^{0.50} \tag{3.39}$$

This equation can be written in terms of d_c and the dimensionless ratios of Table 3.2 as (Bos 1976):

$$Q = C_d\ \frac{A_c}{d_c^2}\ d_c^{2.50}\ \{2g(H_1/d_c - y_c/d_c)\}^{0.50} \tag{3.40}$$

Table 3.2. Ratios for determining the discharge Q of a broad-crested weir and long-throated flume with circular control section (Bos, 1976)

y_c/d_c	$v_c^2/2gd_c$	H_1/d_c	A_c/d_c^2	y_c/H_1	$f(\theta)$	y_c/d_c	$v_c^2/2gd_c$	H_1/d_c	A_c/d_c^2	y_c/H_1	$f(\theta)$
.01	.0033	.0133	.0013	.752	0.0001	.51	.2014	.7114	.4027	.717	0.2556
.02	.0067	.0267	.0037	.749	0.0004	.52	.2065	.7265	.4127	.716	0.2652
.03	.0101	.0401	.0069	.749	0.0010	.53	.2117	.7417	.4227	.715	0.2750
.04	.0134	.0534	.0105	.749	0.0017	.54	.2170	.7570	.4327	.713	0.2851
.05	.0168	.0668	.0147	.748	0.0027	.55	.2224	.7724	.4426	.712	0.2952
.06	.0203	.0803	.0192	.748	0.0039	.56	.2279	.7879	.4526	.711	0.3056
.07	.0237	.0937	.0242	.747	0.0053	.57	.2335	.8035	.4625	.709	0.3161
.08	.0271	.1071	.0294	.747	0.0068	.58	.2393	.8193	.4724	.708	0.3268
.09	.0306	.1206	.0350	.746	0.0087	.59	.2451	.8351	.4822	.707	0.3376
.10	.0341	.1341	.0409	.746	0.0107	.60	.2511	.8511	.4920	.705	0.3487
.11	.0376	.1476	.0470	.745	0.0129	.61	.2572	.8672	.5018	.703	0.3599
.12	.0411	.1611	.0534	.745	0.0153	.62	.2635	.8835	.5115	.702	0.3713
.13	.0446	.1746	.0600	.745	0.0179	.63	.2699	.8999	.5212	.700	0.3829
.14	.0482	.1882	.0688	.744	0.0214	.64	.2765	.9165	.5308	.698	0.3947
.15	.0517	.2017	.0739	.744	0.0238	.65	.2833	.9333	.5404	.696	0.4068
.16	.0553	.2153	.0811	.743	0.0270	.66	.2902	.9502	.5499	.695	0.4189
.17	.0589	.2289	.0885	.743	0.0304	.67	.2974	.9674	.5594	.693	0.4314
.18	.0626	.2426	.0961	.742	0.0340	.68	.3048	.9848	.5687	.691	0.4440
.19	.0662	.2562	.1039	.742	0.0378	.69	.3125	1.0025	.5780	.688	0.4569
.20	.0699	.2699	.1118	.741	0.0418	.70	.3204	1.0204	.5872	.686	0.4701
.21	.0736	.2836	.1199	.740	0.0460	.71	.3286	1.0386	.5964	.684	0.4835
.22	.0773	.2973	.1281	.740	0.0504	.72	.3371	1.0571	.6054	.681	0.4971
.23	.0811	.3111	.1365	.739	0.0550	.73	.3459	1.0759	.6143	.679	0.5109
.24	.0848	.3248	.1449	.739	0.0597	.74	.3552	1.0952	.6231	.676	0.5252
.25	.0887	.3387	.1535	.738	0.0647	.75	.3648	1.1148	.6319	.673	0.5397
.26	.0925	.3525	.1623	.738	0.0698	.76	.3749	1.1349	.6405	.670	0.5546
.27	.0963	.3663	.1711	.737	0.0751	.77	.3855	1.1555	.6489	.666	0.5698
.28	.1002	.3802	.1800	.736	0.0806	.78	.3967	1.1767	.6573	.663	0.5855
.29	.1042	.3942	.1890	.736	0.0863	.79	.4085	1.1985	.6655	.659	0.6015
.30	.1081	.4081	.1982	.735	0.0922	.80	.4210	1.2210	.6735	.655	0.6180
.31	.1121	.4221	.2074	.734	0.0982	.81	.4343	1.2443	.6815	.651	0.6351
.32	.1161	.4361	.2167	.734	0.1044	.82	.4485	1.2685	.6893	.646	0.6528
.33	.1202	.4502	.2260	.733	0.1108	.83	.4638	1.2938	.6969	.641	0.6712
.34	.1243	.4643	.2355	.732	0.1174	.84	.4803	1.3203	.7043	.636	0.6903
.35	.1284	.4784	.2450	.732	0.1289	.85	.4982	1.3482	.7115	.630	0.7102
.36	.1326	.4926	.2546	.731	0.1311	.86	.5177	1.3777	.7186	.624	0.7312
.37	.1368	.5068	.2642	.730	0.1382	.87	.5392	1.4092	.7254	.617	0.7533
.38	.1411	.5211	.2739	.729	0.1455	.88	.5632	1.4432	.7320	.610	0.7769
.39	.1454	.5354	.2836	.728	0.1529	.89	.5900	1.4800	.7384	.601	0.8021
.40	.1497	.5497	.2934	.728	0.1605	.90	.6204	1.5204	.7445	.592	0.8293
.41	.1541	.5641	.3032	.727	0.1683	.91	.6555	1.5655	.7504	.581	0.8592
.42	.1586	.5786	.3130	.726	0.1763	.92	.6966	1.6166	.7560	.569	0.8923
.43	.1631	.5931	.3229	.725	0.1844	.93	.7459	1.6759	.7612	.555	0.9297
.44	.1676	.6076	.3328	.724	0.1927	.94	.8065	1.7465	.7662	.538	0.9731
.45	.1723	.6223	.3428	.723	0.2012	.95	.8841	1.8341	.7707	.518	1.0248
.46	.1769	.6369	.3527	.722	0.2098						
.47	.1817	.6517	.3627	.721	0.2186						
.48	.1865	.6665	.3727	.720	0.2276						
.49	.1914	.6814	.3827	.719	0.2368						
.50	.1964	.6964	.3927	.718	0.2461						

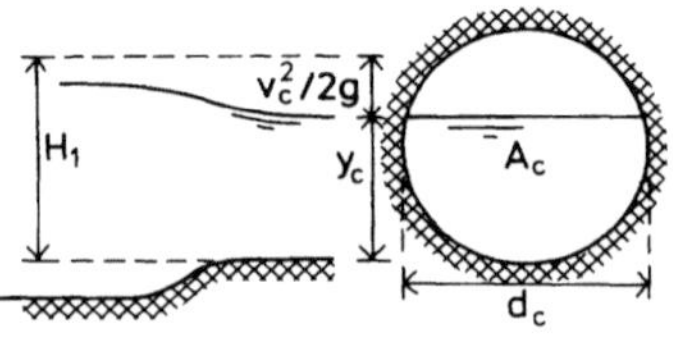

Substitution of Eqs. 3.34, 3.36, and 3.38 into Eq. 3.40 yields:

$$Q = C_d d_c^{2.50} g^{0.50} \{f(\Theta)\} \tag{3.41}$$

where $f(\Theta) = \frac{(\Theta - \sin\Theta)^{1.5}}{8(8\sin\frac{\Theta}{2})^{0.5}}$ is a shape factor for the control section.

If d_c is known and H_1 is set to a given value, the related value of $f(\Theta)$ can be read from Table 3.2. Substitution of this value and the C_d value into Eq. 3.40 yields the discharge Q. The iterative procedure of Section 3.3.5 should be used to transform this H_1-Q relationship into an h_1-Q relationship.
Table 3.2 also contains columns presenting dimensionless values for the velocity head, water depth, and related area of flow. These data will be used in Section 3.3.9 and in Chapter 4.

3.3.8 U-shaped control section

For measuring structures with a U-shaped control section (Figure 3.10), two head-discharge equations must be used: the first for the conditions where flow is confined within the circular part of the control section, and the second, at higher stages, where the presence of the vertical side walls has to be taken into account. The first equation is:

$$Q = C_d \frac{A_c}{d_c^2} d_c^{2.50} \{2g(H_1/d_c - y_c/d_c)\}^{0.50} \qquad \text{(Eq. 3.40)}$$

This equation is valid if $y_c < \frac{1}{2}d_c$ or if $\Theta < \pi$.
Entering Table 3.2 with this limitation shows the related ratio H_1/d_c = 0.6964. Hence, the use of Eq. 3.40 for the U-shaped control is limited by $H_1 < 0.70\ d_c$.

If $H_1 > 0.70\ d_1$, the second equation should be used.
It was derived by Bos in 1977 as follows:

If $y_c > \frac{1}{2}d_c$, we can write:

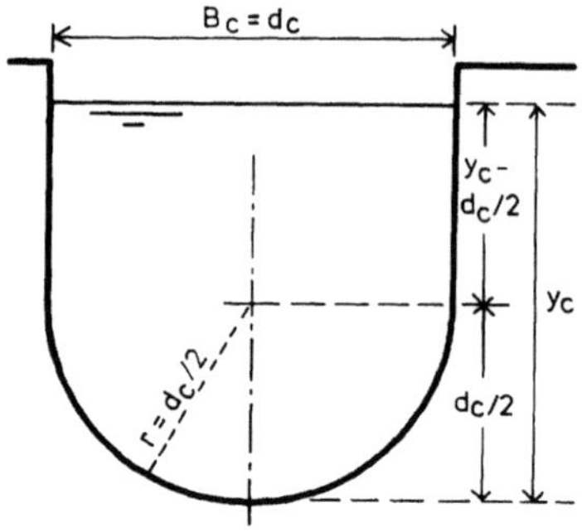

Figure 3.10. Dimensions of U-shaped control section

$$A_c = \frac{1}{8}\pi d_c^{\ 2} + d_c(y_c - \frac{1}{2}d_c) \tag{3.42}$$

and:

$$B_c = d_c \tag{3.43}$$

Substitution of this information into Eq. 3.13 gives for the critical depth:

$$y_c = \frac{2}{3} H_1 + 0.0358d_c \tag{3.44}$$

Combining Eqs. 3.12, 42, and 44, and adding the C_d and C_v coefficients gives:

$$Q = C_d C_v \frac{2}{3} (\frac{2}{3}g)^{0.50} d_c(h_1 - 0.1073d_c)^{1.50} \tag{3.45}$$

A comparison with Eq. 3.18 shows that $h_1 - 0.1073d_c$ in Eq. 3.45 is the upstream head with respect to the equivalent horizontal sill in the control section.

3.3.9 Truncated circular control section

For a structure with a circular control section that is truncated on the bottom (Figure 3.11), we may write the following relationships:

$$A_s = d_c^2 \; (\phi - \sin\phi)/8 \qquad (3.46)$$

$$A_t = d_c^2 \; (\Theta - \sin\Theta)/8 \qquad (3.47)$$

$$B_c = d_c \sin\frac{\Theta}{2} \qquad (3.48)$$

$$b_c = d_c \sin\frac{\phi}{2} \qquad (3.49)$$

$$p_c = d_c \; (1 - \cos\frac{\phi}{2})/2 \qquad (3.50)$$

and:

$$y_c = d_c \; (\cos\frac{\phi}{2} - \cos\frac{\Theta}{2})/2 \qquad (3.51)$$

where:

A_s = area of sill;
A_t = total area of sill and wetted flow area at the control section;
b_c = sill width;
p_c = sill height at the control section.

If the pipe is horizontal, note that $p_1 = p_c$ (see Fig. 3.3).

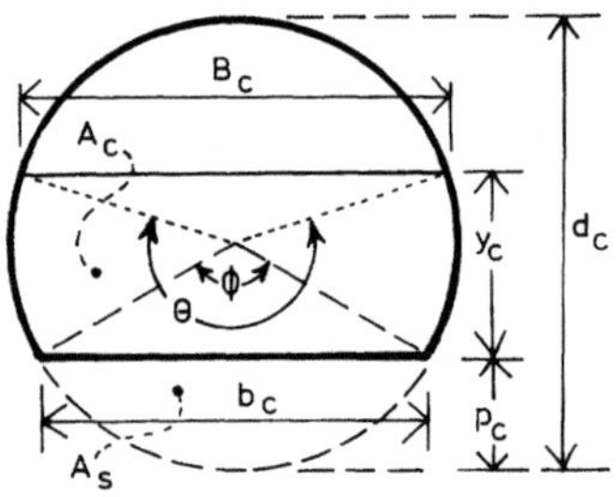

Figure 3.11. Dimensions of circular control section with truncated bottom

The wetted area at the control section, A_c, is found from

$$A_c = A_t - A_s = d_c^2 (\Theta - \phi + \sin\phi - \sin\Theta)/8 \tag{3.52}$$

Substituting the above A_c and B_c values into Eq. 3.8 gives:

$$\frac{v_c^2}{2g} = \frac{A_c}{2B_c} = \frac{d_c^2 (\Theta - \phi + \sin\phi - \sin\Theta)}{16 \sin\frac{\Theta}{2}} \tag{3.53}$$

which, in combination with Eq. 3.13, yields:

$$\frac{H_1}{d_c} - \frac{y_c}{d_c} = \frac{\Theta + \phi + \sin\phi - \sin\Theta}{16 \sin\frac{\Theta}{2}} \tag{3.54}$$

Substituting Eqs. 3.52 and 3.54 into Eq. 3.12 and introducing a C_d value gives (Clemmens, Bos, and Replogle 1984):

$$Q = C_d \, d_c^{2.50} g^{0.50} \frac{(\Theta - \phi + \mathrm{sn}\phi - \sin\Theta)^{1.50}}{8(8 \sin \frac{\Theta}{2})^{0.50}} \tag{3.55}$$

or

$$Q = C_d \, d_c^{2.50} g^{0.50} \{f\,(\phi, \Theta)\} \tag{3.56}$$

To solve this equation for a given combination of p_c and d_c (and thus for ϕ through Eq. 3.50), it should be noted that the angle Θ is defined if a y_c value is given in addition (Eq. 3.51). Hence, the angle Θ is defined if $(p_c + y_c)/d_c$ is known. Adding $v_c^2/2gd_c$ (from Eq. 3.52) to

Table 3.3. Shape factors $f(\phi, \Theta)$ of a broad-crested weir in a circular pipe. $C_d = 1.0$, $\alpha_c = 1.0$, $H_1 = H_c$.

$\frac{p_c + H_1}{d_c}$	p_c/d_c = 0.15	0.20	0.25	0.30	0.35	0.40	0.45	0.50
0.16	0.0004							
0.17	0.0011							
0.18	0.0021							
0.19	0.0032							
0.20	0.0045							
0.21	0.0060	0.0004						
0.22	0.0076	0.0012						
0.23	0.0094	0.0023						
0.24	0.0113	0.0036						
0.25	0.0133	0.0050						
0.26	0.0155	0.0066	0.0005					
0.27	0.0177	0.0084	0.0013					
0.28	0.0201	0.0103	0.0025					
0.29	0.0226	0.0124	0.0038					
0.30	0.0252	0.0145	0.0054					
0.31	0.0280	0.0169	0.0071	0.0005				
0.32	0.0308	0.0193	0.0090	0.0014				
0.33	0.0337	0.0219	0.0110	0.0026				
0.34	0.0368	0.0245	0.0132	0.0040				
0.35	0.0399	0.0273	0.0155	0.0057				
0.36	0.0432	0.0302	0.0179	0.0075	0.0005			
0.37	0.0465	0.0332	0.0205	0.0094	0.0015			
0.38	0.0500	0.0363	0.0232	0.0115	0.0027			
0.39	0.0535	0.0396	0.0260	0.0138	0.0042			
0.40	0.0571	0.0429	0.0289	0.0162	0.0059			
0.41	0.0609	0.0463	0.0320	0.0187	0.0077	0.0005		
0.42	0.0647	0.0498	0.0351	0.0214	0.0097	0.0015		
0.43	0.0686	0.0534	0.0383	0.0242	0.0119	0.0028		
0.44	0.0726	0.0571	0.0417	0.0271	0.0143	0.0043		
0.45	0.0767	0.0609	0.0451	0.0301	0.0167	0.0060		
0.46	0.0809	0.0648	0.0487	0.0332	0.0193	0.0079	0.0005	
0.47	0.0851	0.0688	0.0523	0.0365	0.0220	0.0100	0.0015	
0.48	0.0895	0.0729	0.0561	0.0398	0.0249	0.0122	0.0028	
0.49	0.0939	0.0770	0.0599	0.0432	0.0279	0.0145	0.0043	
0.50	0.0984	0.0813	0.0638	0.0468	0.0309	0.0170	0.0061	
0.51	0.1030	0.0856	0.0678	0.0504	0.0341	0.0197	0.0080	0.0005
0.52	0.1076	0.0900	0.0719	0.0541	0.0374	0.0224	0.0101	0.0015
0.53	0.1124	0.0945	0.0761	0.0579	0.0408	0.0253	0.0123	0.0028
0.54	0.1172	0.0990	0.0803	0.0618	0.0443	0.0283	0.0147	0.0044
0.55	0.1221	0.1037	0.0847	0.0658	0.0479	0.0314	0.0172	0.0061
0.56	0.1270	0.1084	0.0891	0.0699	0.0515	0.0346	0.0198	0.0080
0.57	0.1320	0.1132	0.0936	0.0741	0.0553	0.0379	0.0226	0.0101
0.58	0.1372	0.1180	0.0981	0.0783	0.0592	0.0413	0.0255	0.0123
0.59	0.1423	0.1230	0.1028	0.0826	0.0631	0.0448	0.0285	0.0147
0.60	0.1476	0.1280	0.1075	0.0870	0.0671	0.0484	0.0316	0.0172
0.62		0.1382	0.1172	0.0960	0.0754	0.0559	0.0381	0.0225
0.64		0.1486	0.1271	0.1053	0.0840	0.0637	0.0449	0.0283
0.66		0.1593	0.1373	0.1149	0.0929	0.0718	0.0522	0.0346
0.68		0.1703	0.1477	0.1247	0.1020	0.0802	0.0597	0.0412
0.70		0.1815	0.1584	0.1348	0.1114	0.0888	0.0676	0.0481
0.72		0.1929	0.1692	0.1451	0.1211	0.0978	0.0757	0.0554
0.74		0.2045	0.1804	0.1556	0.1310	0.1070	0.0841	0.0629
0.76		0.2163	0.1917	0.1663	0.1411	0.1164	0.0928	0.0707
0.78		0.2283	0.2031	0.1773	0.1514	0.1260	0.1016	0.0788
0.80		0.2405	0.2148	0.1884	0.1618	0.1358	0.1107	0.0870
0.82		0.2528	0.2267	0.1997	0.1725	0.1458	0.1200	0.0955
0.84		0.2653	0.2386	0.2111	0.1833	0.1559	0.1294	0.1042
0.86		0.2780	0.2508	0.2227	0.1943	0.1662	0.1390	0.1130
0.88		0.2907	0.2630	0.2344	0.2054	0.1767	0.1487	0.1220
0.90		0.3036	0.2754	0.2462	0.2166	0.1872	0.1586	0.1311
0.92		0.3166	0.2879	0.2581	0.2279	0.1979	0.1686	0.1404
0.94		0.3297	0.3005	0.2701	0.2394	0.2087		
0.96		0.3428	0.3131	0.2823	0.2509			
0.98		0.3561	0.3259	0.2944				
1.00		0.3694	0.3387					
1.02		0.3827						
1.04		0.3961						

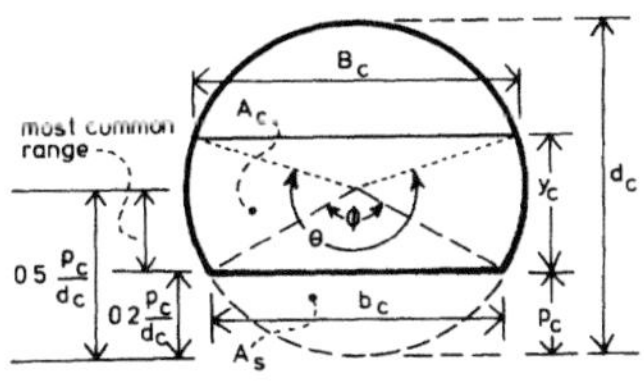

this ratio yields:

$$\frac{H_1 + p_c}{d_c} = \frac{8 \sin\frac{\Theta}{2} - 5 \sin\Theta + \Theta + \phi + \sin\phi}{16 \sin\frac{\Theta}{2}} \qquad (3.57)$$

Thus for given values of relative sill height p_c/d_c and of $(H_1 + p_c)/d_c$, the $f(\phi, \Theta)$ can be solved. Values for the shape factor $f(\phi, \Theta)$ have been computed and are given in Table 3.3.

For a structure of known dimensions, p_c and d_c, and a given H_1 value, the related dimensionless value can be obtained from Table 3.3, and Q can then be calculated from Eq. 3.56.
The resulting Q-H_1 relationship can be transformed into a Q-h_1 table or curve by using the method of Section 3.3.5. Then for any dimensionless water depth, y_1/d_1, a corresponding value of A_1/d_1^2 can be found in Table 3.2. In using Table 3.3, it should be noted that the velocity head is included and therefore h_1 should not be confused with H_1.

3.3.10 Summary of equations

In the above sections, head-discharge equations were derived for a variety of shapes of the control section. All these equations are based upon Eqs. 3.8, 3.12, and 3.13, while the assumptions that underlie the C_d and C_v coefficients are identical in all cases. The derived equations are summarized in Figure 3.12. The y_c values are also summarized because they are needed to estimate the modular limit (Chapter 4).

3.4 EQUIVALENT SHAPES OF CONTROL SECTIONS

An important property of the broad-crested or long-throated structure follows directly from a combination of the Eqs. 3.12 and 3.13. The resulting equation:

$$Q_i = \sqrt{g\, A_c^{\,3}/B_c} \qquad (3.58)$$

SHAPE OF CONTROL SECTION	HEAD–DISCHARGE EQ. TO BE USED	HOW TO FIND THE y_c - VALUE	SOURCE OF BASIC EQUATION
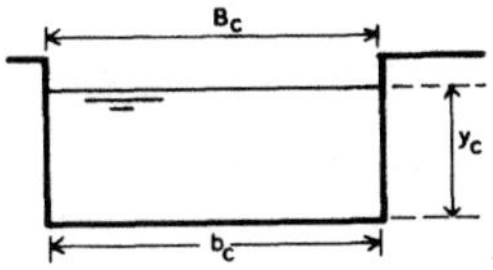	$Q = C_d C_v \frac{2}{3} (\frac{2}{3} g)^{1/2} b_c h_1^{3/2}$	$y_c = \frac{2}{3} H_1$	Belanger, 1849
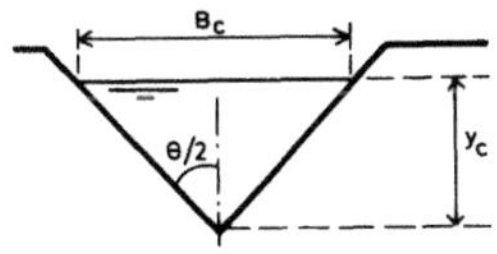	$Q = C_d C_v \frac{16}{25} (\frac{2}{5} g)^{1/2} \tan\frac{\theta}{2}\, h_1^{5/2}$	$y_c = \frac{4}{5} H_1$	Jameson, 1925
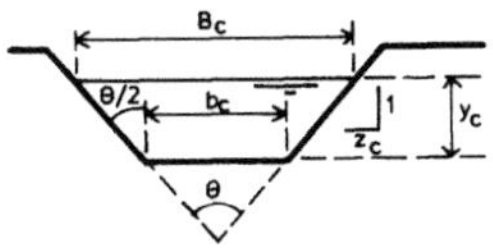	$Q = C_d [b_c y_c + z_c y_c^2][2g(H_1 - y_c)]^{1/2}$	Use Table 3.1	Bos, 1976
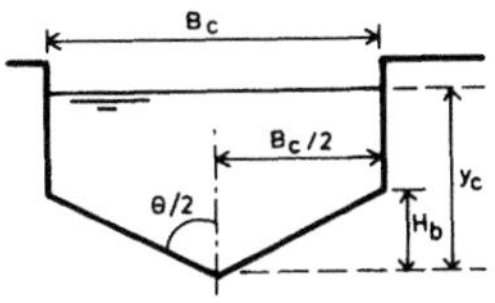	If $H_1 \leq 1.25\, H_b$ $Q = C_d C_v \frac{16}{25} (\frac{2}{5} g)^{1/2} \tan\frac{\theta}{2}\, h_1^{5/2}$	$y_c = \frac{4}{5} H_1$	Jameson, 1925
	If $H_1 \geq 1.25\, H_b$ $Q = C_d C_v \frac{2}{3} (\frac{2}{3} g)^{1/2} B_c (h_1 - \frac{1}{2} H_b)^{3/2}$	$y_c = \frac{2}{3} H_1 + \frac{1}{6} H_b$	Bos, 1976
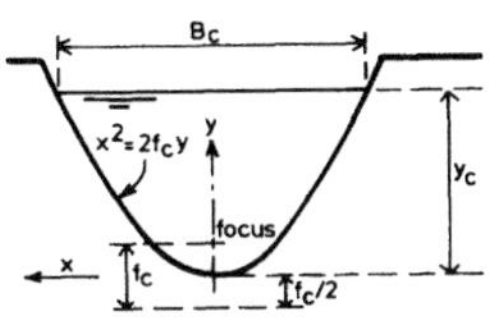	$Q = C_d C_v (\frac{3}{4} f_c g)^{1/2} h_1^2$	$y_c = \frac{3}{4} H_1$	Jameson, 1925
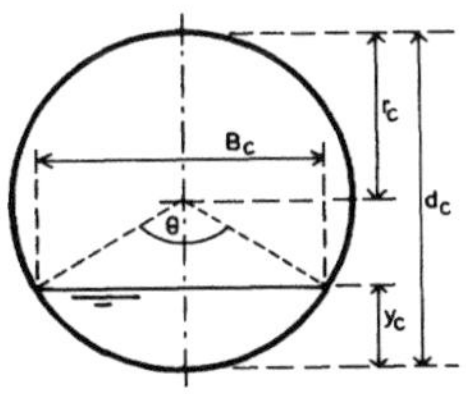	$Q = C_d d_c^{5/2} \sqrt{g}\, [f(\theta)]$ use table 3.2 to find $f(\theta)$	Use Table 3.2	Bos, 1976
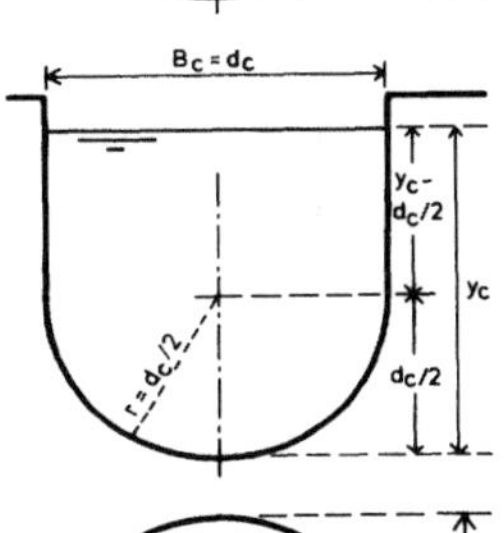	If $H_1 \leq 0.70\, d_c$ $Q = C_d d_c^{5/2} \sqrt{g}\, [f(\theta)]$ use table 3.2 to find $f(\theta)$	Use Table	Bos, 1976
	If $H_1 \geq 0.70\, d_c$ $Q = C_d C_v \frac{2}{3} (\frac{2}{3} g)^{1/2} d_c (h_1 - 0.1073 d_c)^{3/2}$	$y_c = \frac{2}{3} H_1 + 0.0358 d_c$	Bos, 1977
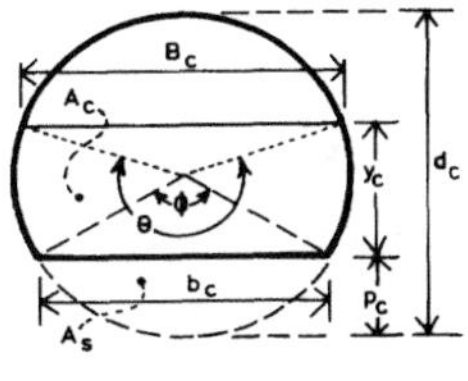	$Q = C_d d_c^{5/2} \sqrt{g}\, [f(\phi, \theta)]$ use table 3.3 to find $f(\phi, \theta)$	y_c is variable	Clemmens, Bos and Replogle, 1984

Figure 3.12. Summary of equations

shows that the (ideal) discharge through a control section is determined by two variables only: A_c and B_c. In other words, all control shapes that flow with the same water surface width, B_c, and have an arbitrarily shaped wetted area with the same A_c value, discharge at the same rate. This property, which was illustrated with the derivation of Eqs. 3.27 and 3.45, can be used for other control shapes with composite cross sections. The following two examples may further illustrate this.

Photo 3.1. Long-throated flume with complex control section

The left-hand side of Figure 3.13 shows the designs for the control section of two long-throated flumes. In both designs, the head under which the lower flows are to be measured is increased by narrowing the bottom to b_c = 0.50 m.

If the critical water depth at the complex control section does not exceed 0.25 m, the head-discharge relationship of the identical trapezoïdal parts of the controls is given by Eq. 3.31. The limit of application of this equation can be derived from Eq. 3.29: Substitution of b_c = 0.50 m, y_c = 0.25 m, and z_c = 2.0 into that equation yields H_1 = 0.333 m as the matching upper limit for H_1.

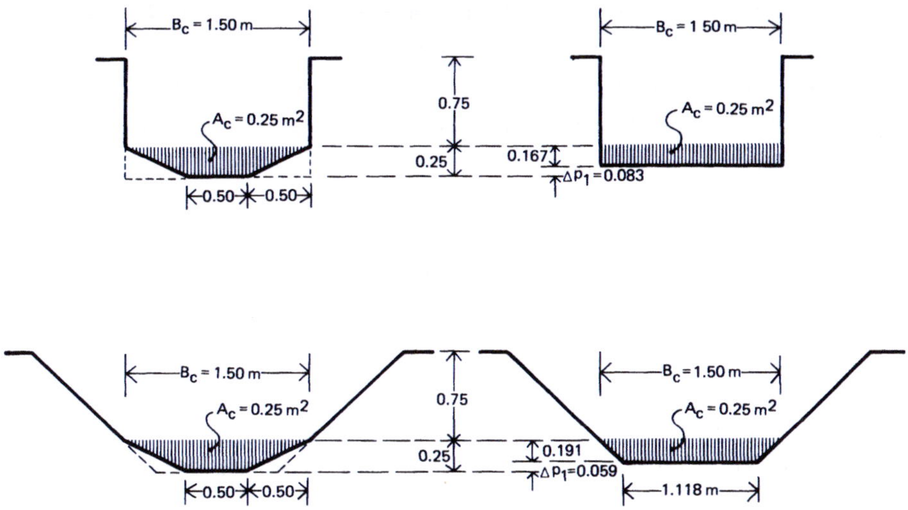

Figure 3.13. Similarity in shape of control sections with equal discharge

If H_1 > 0.333 m, no head-discharge equation is available for the two complex control sections. The lower trapezoids, however, can be replaced by, respectively, a rectangle and a trapezoid with B_c = 1.50 m and A_c = 0.25 m^2. The dimensions of these two areas are shown in the right-hand side of Figure 3.13. As can be seen, the equivalent singular control shapes have an imaginary invert at 0.083 m and 0.059 m respectively above the real sill level. These values must be subtracted from the real H_1 values of the left-hand side controls if applied to the head-discharge equations of the equivalent singular control sections.

The procedure of obtaining a rating curve for a complex control shape is illustrated below, with the control shape at the top left of Figure 3.13 serving as an example:

1. Determine the y_c value for which the lower part of the control section flows full (y_c = 0.25 m in example).
2. Calculate the H_1 value which corresponds with this y_c value; Eq. 3.29 yields H_1 = 0.333 m for the original control shape.

3. Select the relevant head-discharge equation from Figure 3.12 and calculate a rating curve for the lower part of the control section.
4. Determine the area of the lower part of the control section $A_c = 0.25\ m^2$ and calculate the dimensions of the 'equivalent rectangular area'.
5. Calculate the imaginary raise, Δp_1, of the sill-reference level of the 'equivalent singular control' ($\Delta p_1 = 0.083$ m in example).
6. Select the head-discharge equation for the 'equivalent rectangular control'. Calculate the related rating curve. Correct the heads of this rating curve with the Δp_1 of Step 5 (add $\Delta p_1 = 0.083$ m to imaginary head).
7. Convert the Q-H_1 rating into a Q-h_1 rating (Section 3.3.5).
8. Summarize the rating curves as illustrated in Figure 3.14.

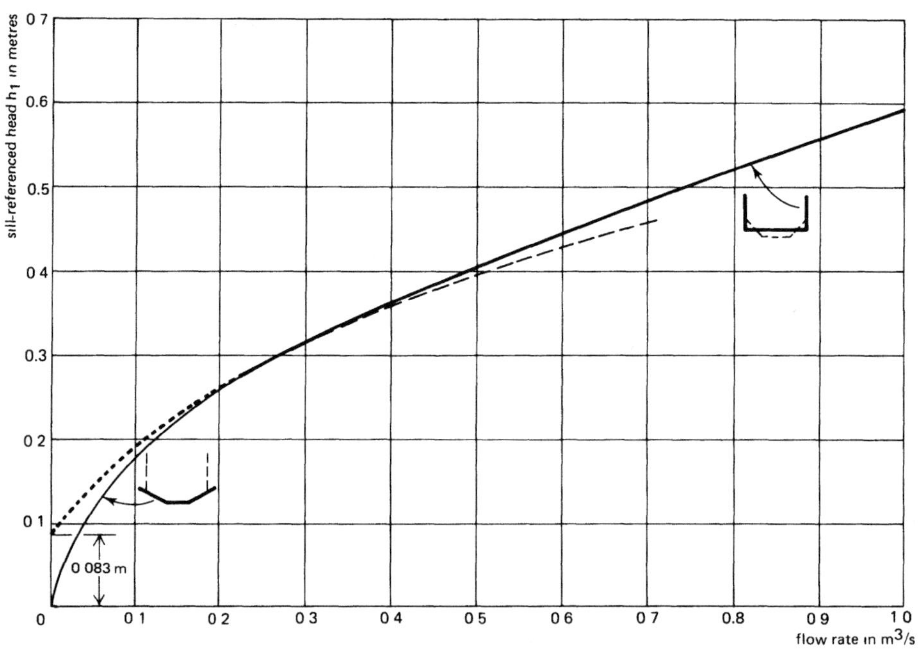

Figure 3.14. Rating curves for a complex control shape

3.5 DISCHARGE COEFFICIENT, C_d

3.5.1 Physical meaning

As stated when the discharge coefficient was being introduced in Eq. 3.17, its numerical value follows from:

$$C_d = Q/Q_i \tag{3.59}$$

C_d corrects for the actual deviation from the following three idealizing assumptions:

* Absence of energy losses between gauging station and control section;
* Straight and parallel streamlines at gauging station and at control section;
* Uniform velocity distribution at the gauging station and control section.

3.5.2 C_d values

As mentioned in Section 3.5.1, the value of the discharge coefficient is mainly influenced by energy loss due to friction, and by an increase of velocities due to streamline curvature. Both phenomena can be related to the dimensionless ratio H_1/L. For heads that are low with respect to L, energy loss due to friction upstream of the control section is significant with respect to H_1, causing C_d to decrease with decreasing H_1/L. For high H_1/L ratios, the downward streamline curvature causes C_d to rise.

Laboratory data on 105 broad-crested weirs and long-throated flumes from 29 different research papers (for list of authors, see legend of Figure 3.15) were used to calculate the C_d value as a function of the H_1/L ratio. The resulting 1 395 points are plotted in Figure 3.15, which shows the relation between C_d and H_1/L for the wide variety of weirs and flumes that were tested. The dimensions and construction materials of these structures differed considerably: the shape of the control section ranged from a narrow ($\Theta = 30°$) triangle to a wide ($b_c = 16.45$ m) trapezoid; the throat length ranged from 0.04 m to 3.65 m; the converging transition was made either of curved surfaces ($r_{bottom} = 0.2\ H_{1max}$) or of flat surfaces with a contraction ratio as low as 1-to-4; approach velocity conditions ranged from two-dimensional to three-dimensional, while at Q_{max} the Froude number in the approach channel ranged from 0.1 to 0.5; the downstream diverging transition

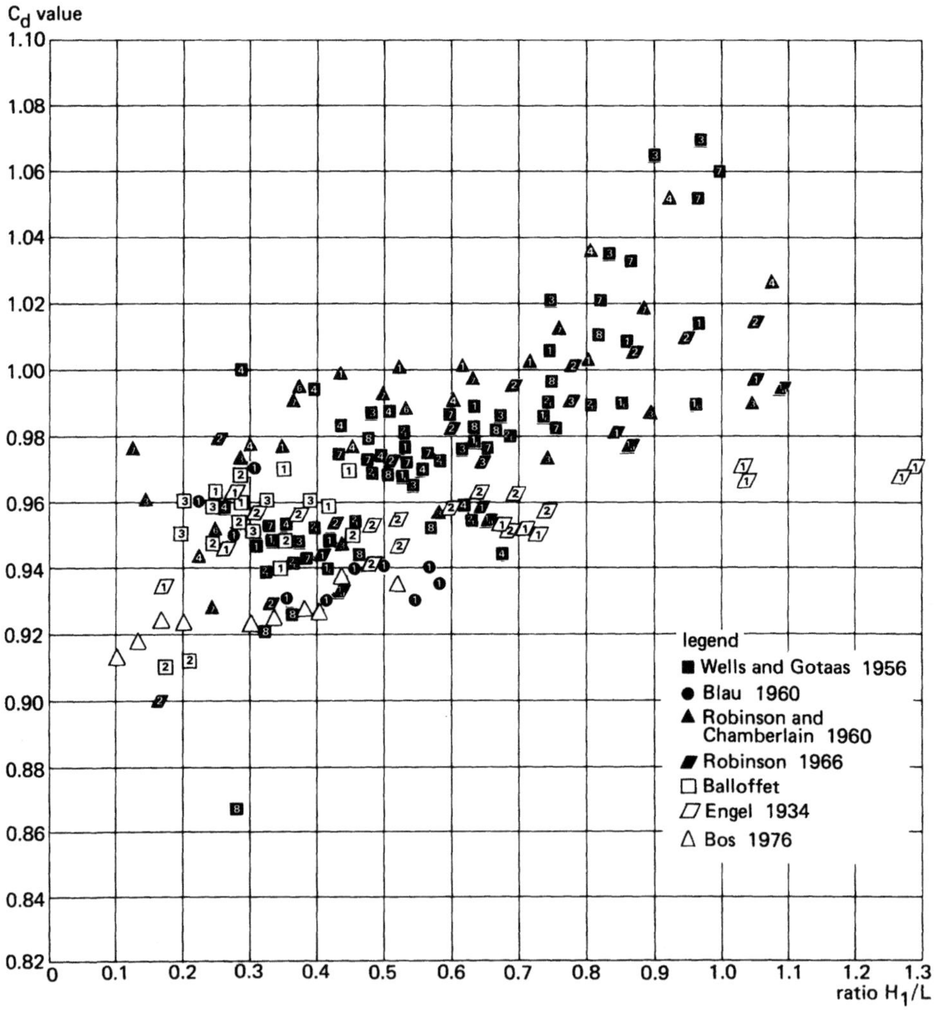

Figure 3.15. Laboratory tested C_d value plotted versus the ratio H_1/L

ranged from very gradual (10-to-1) to fully truncated (0-to-1); flow towards the control section was accelerated by side contraction only, by a bottom sill only, or by a combination of the two; structures were built of glass, brass, PVC, sheet steel, wood, and concrete. The results of the following analysis thus apply to smoothly finished broad-crested weirs and long-throated flumes.

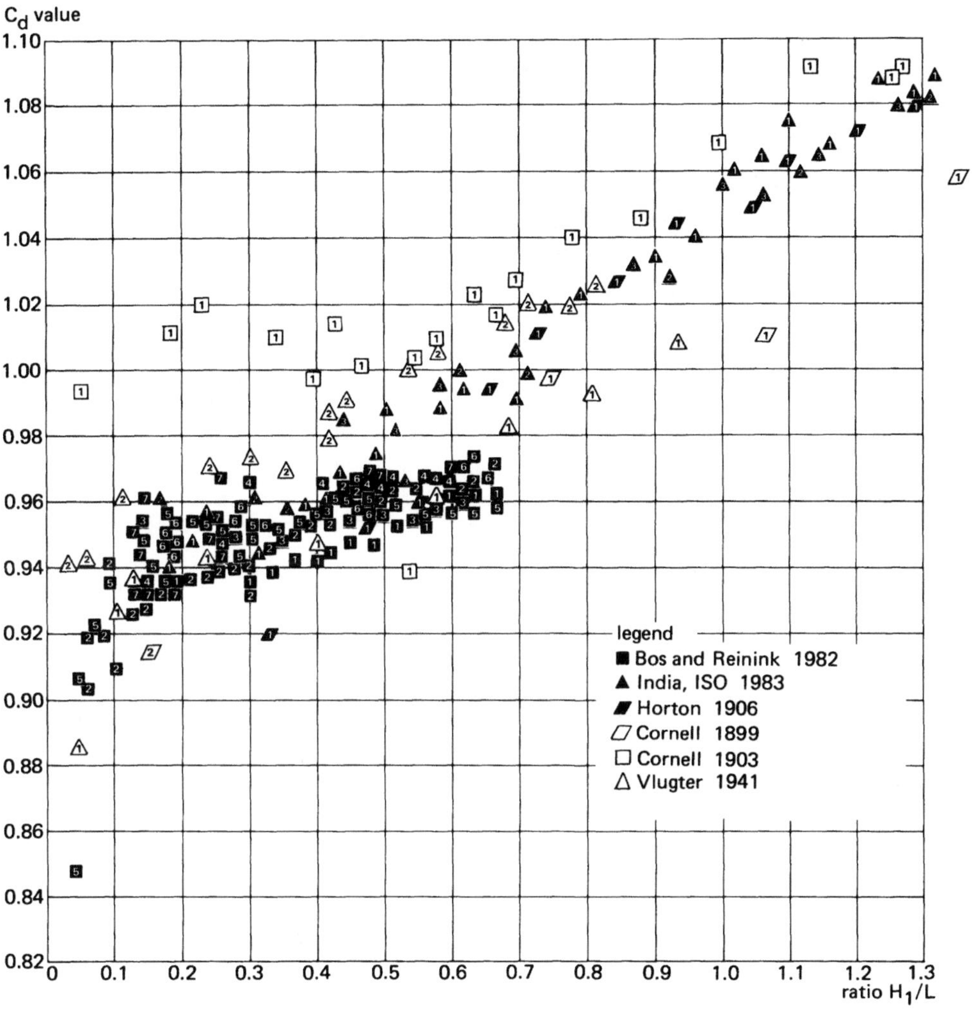

Figure 3.15 (cont.)

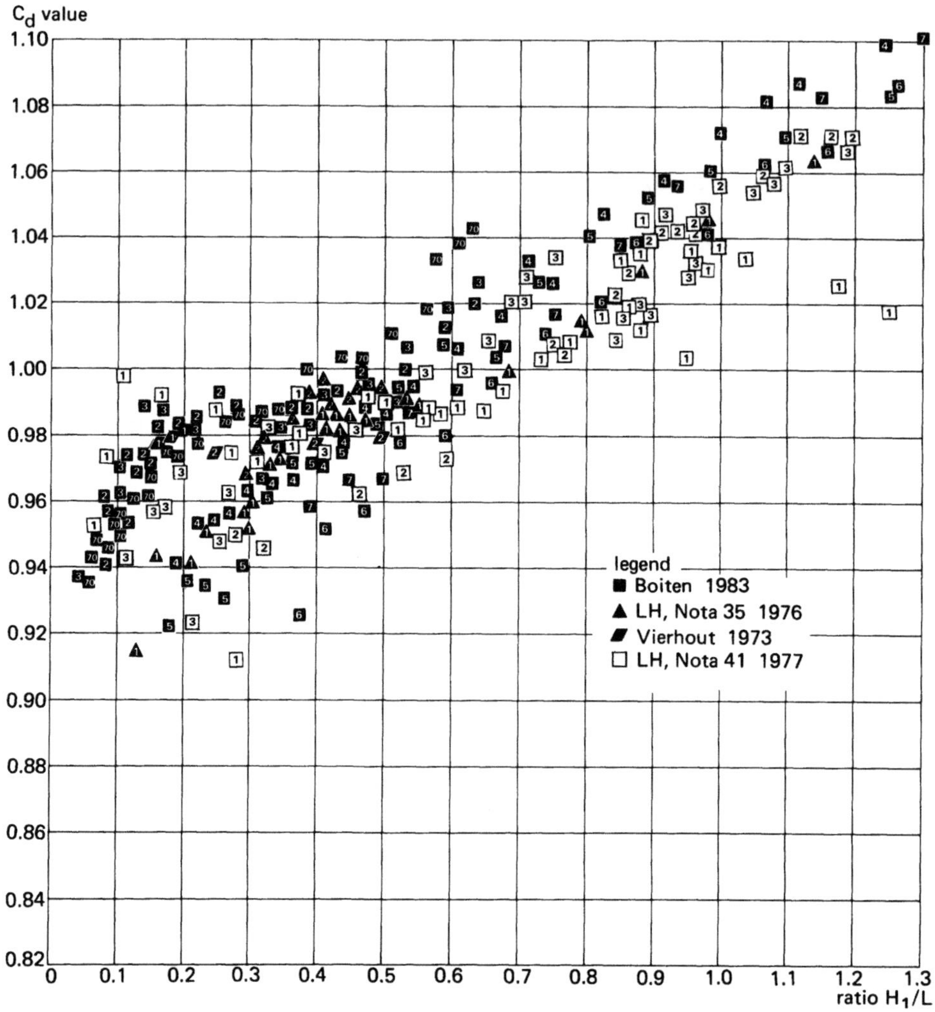

Figure 3.15. (cont.)

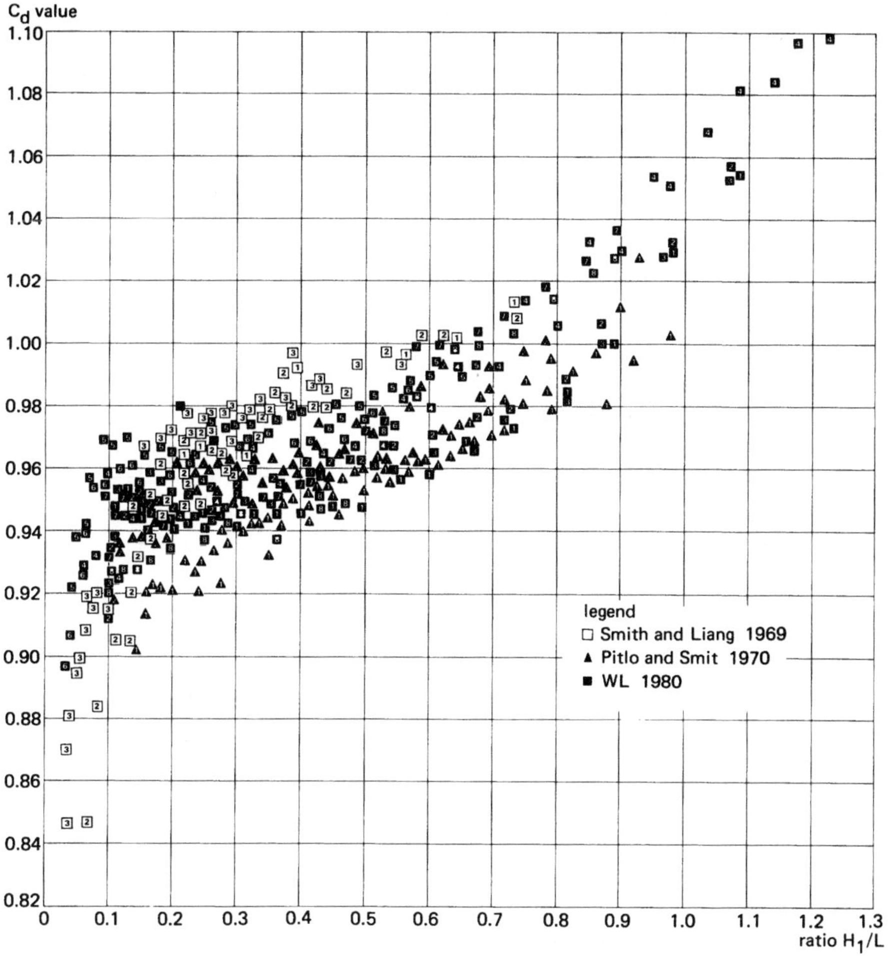

Figure 3.15. (cont.)

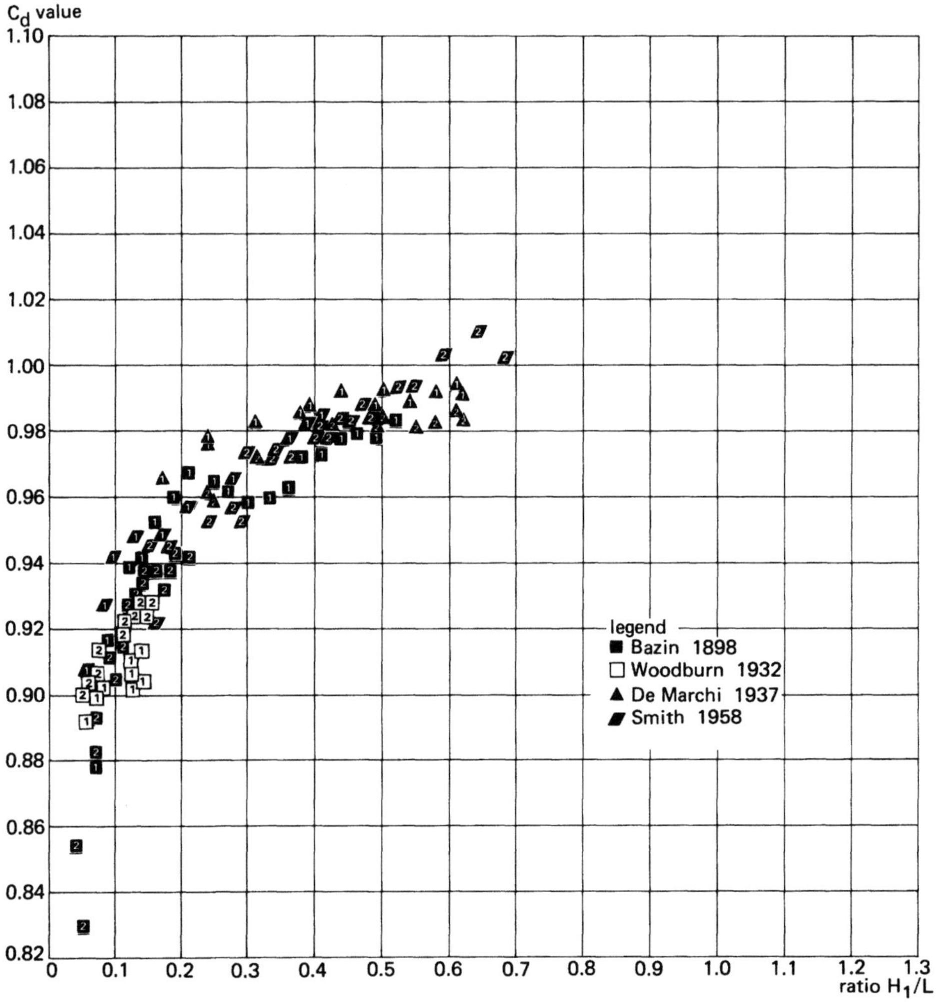

Figure 3.15. (cont.)

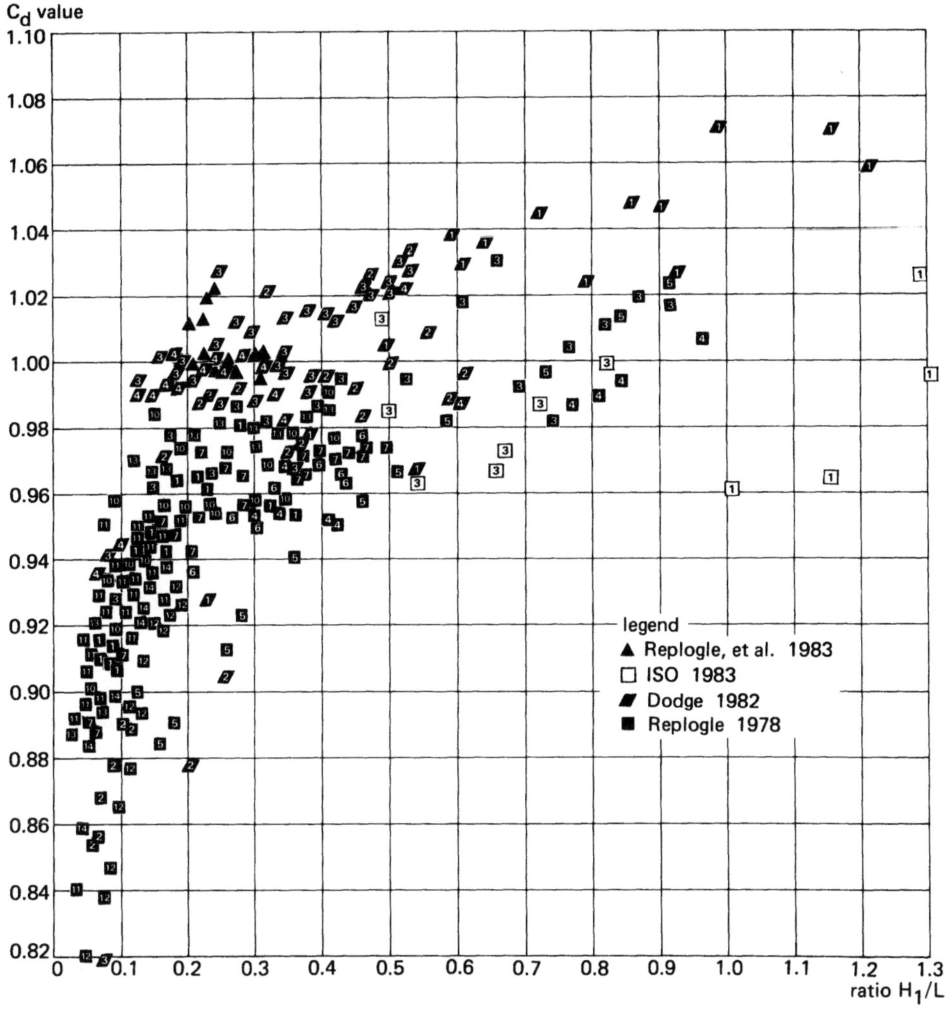

Figure 3.15. (cont.)

The average value of C_d, and its standard deviation, for classified values of H_1/L are listed in Table 3.4. The last column of this table also gives the percentage error of this average C_d value (95% confidence level). Both the average C_d values (0) and the related 95% confidence limits are presented in Figure 3.16.

Table 3.4. Average C_d values and standard deviation for various H_1/L ratios

Ratio H_1/L	Number of data points, n	Average C_d value	Standard deviation, s	95% confidence limit in per cent of C_d
0- . 5	25	.886	.040	9.3
. 5- .10	91	.917	.031	6.7
.10- .15	119	.939	.024	5.0
.15- .25	212	.954	.023	4.7
.25- .35	199	.962	.022	4.5
.35- .45	186	.969	.019	3.8
.45- .55	158	.976	.021	4.2
.55- .65	122	.984	.022	4.4
.65- .75	87	.993	.023	4.6
.75- .85	51	1.009	.018	3.6
.85- .95	52	1.025	.020	3.9
.95-1.05	39	1.034	.030	5.9
1.05-1.15	25	1.061	.023	4.5
1.15-1.25	16	1.064	.035	7.0
1.25-1.35	13	1.056	.048	9.8
	Σn = 1395			

When the data on C_d were being processed, it was assumed that for a given H_1/L ratio the coefficient data were normally distributed with an average value C_d and standard deviation s. As stated in Section 3.5.1, however, energy loss due to friction in the zone of acceleration and streamline curvature at the control section influence the C_d value. For the same area of flow of the control section, the wetted perimeter in a narrow control section is greater than the wetted perimeter in a wide control section. Consequently, friction losses are also greater, and a narrow control section has a lower discharge coefficient than the wide control section. Further, because streamline curvature at the control section can develop better above weirs with a high p_2 value than above structures where the tailwater channel bottom deflects the jet upwards, the discharge coefficient of a high weir (p_2/H_1 > 0.75; Bos 1976) will be higher than that of a flat-bottomed flume.

Corresponding clusters of C_d values for various types of structures are indicated in Figure 3.16. For general practice in water management,

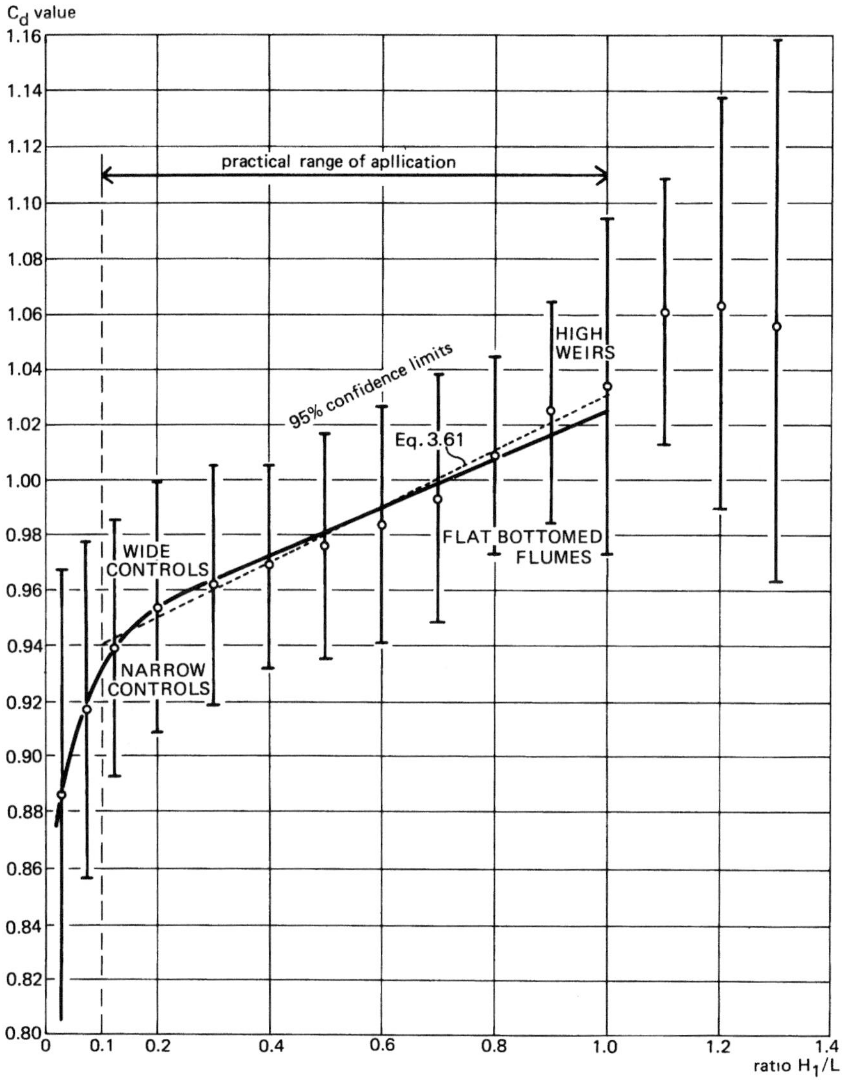

Figure 3.16. C_d values as a function of H_1/L for broad-crested weirs and long-throated flumes of all shapes and sizes

however, it is recommended that these systematic differences be ignored, and that the average C_d versus H_1/L curve of Figure 3.16 be used instead.

To limit the effect of roughness of the weir crest at low H_1/L values and the effect of variable streamline curvature at high H_1/L values, the following practical limits of application are recommended:

$$0.1 < H_1/L < 1.0 \qquad (3.60)$$

It should be noted that this practical range of validity is wider than the theoretical range of Eq. 3.9.
The C_d versus H_1/L line, within the practical limits of application shown in Figure 3.16, closely corresponds with the following equation:

$$C_d = 0.93 + 0.10\ H_1/L \qquad (3.61)$$

The error in the discharge coefficient of Figure 3.16 or Eq. 3.61 was determined from the 95% confidence limits. For a given value of the ratio H_1/L, this coefficient error is:

$$X_c = \pm\ (\ 3\left|H_1/L - 0.55\right|^{1.5} + 4)\ \text{percent} \qquad (3.62)$$

In some special cases, such as a water balance study, accurate measurements in a wider range of flow ratios may be required. The measuring structure must then be calibrated either in the field or in a laboratory scale model.

3.5.3 Influence of transition shapes on C_d

As mentioned earlier, the C_d versus H_1/L relationship of Figure 3.16 is valid for all broad-crested weirs and long-throated flumes provided that there is no flow separation at the leading edge of their crest or throat. Such flow separation can be avoided by properly shaped converging transitions. As already mentioned in Section 2.3.1, the main function of a converging transition is to guide the flow into the control section without causing flow separation to occur.

To study the effect of differently shaped converging transitions on the C_d value, tests were conducted (Bos and Reinink 1981). The data points are plotted in Figure 3.17.

Tests A1 and A5 were run on the model described in Section 4.3.1. This was a trapezoidal-throated flume with a transition in which the bottom converged 2-to-1 and the side slopes 1.50-to-1. The transition started at 0.30 m upstream of the throat entrance.

In Tests B1 and B5, the transition was changed to start at 0.45 m upstream of the throat entrance and the bottom slope was changed to 3-to-1. The side slopes converged 2.25-to-1.

In Test C6, the sides converged as in the B1 and B5 tests, but the bottom of the transition was flat.

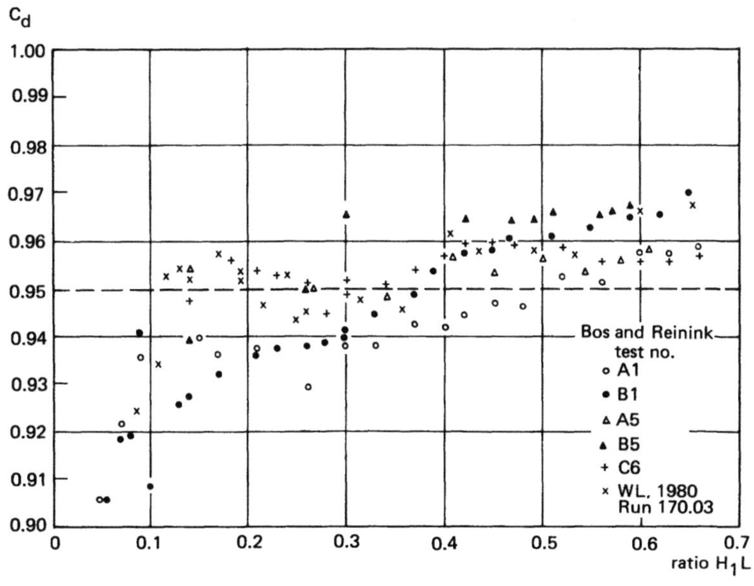

Figure 3.17. Influence of shape of converging transition on value of C_d (WL 1980; Bos and Reinink 1981)

Figure 3.17. also plots the data points of a test conducted at the Hydraulic Laboratory, Delft (WL 1980, test run 170.03). This test was run on a broad-crested weir with bottom contraction only and a rounded converging transition with a radius of 0.2 H_{1max} (r = 0.20 m).
On the basis of the data of Figure 3.17, there would seem to be no practical difference between the effects of a plane transitions (2-to-1 slopes of A1 and A5) and a rounded transition (r = 0.2 H_{1max} of WL 1980).

For ratios of H_1/L greater than 0.5, streamlines become increasingly curved. In flat-bottomed flumes, $p_1 = p_2 = 0$, the discharging jet is supported by the bottom of the downstream transition; the streamline curvature is thus less than in flumes with elevated throats. This difference in C_d values is shown in Figure 3.18 for H_1/L values of 0.45 and 0.6 (Bos and Reinink 1981).

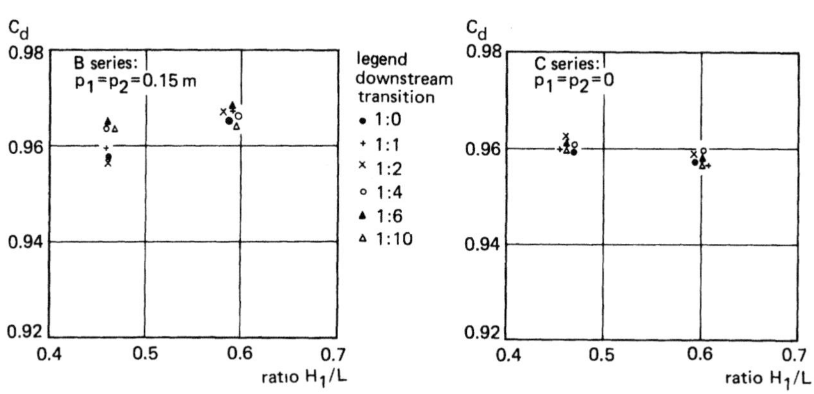

Figure 3.18. Influence of shape of downstream transition on the value of C_d

The flare angle of the downstream transition seems to have no significant influence on C_d. The coefficient error of Eq. 3.62 also includes the effect of the throat elevation p_2, at values of H_1/L up to 1.0.

3.6 APPROACH VELOCITY COEFFICIENT, C_v

3.6.1 Physical meaning

As stated in Section 3.3.2, the approach velocity coefficient corrects for the use of h_1 instead of H_1 in the head-discharge equation, and thus for neglecting $\alpha_1 v_1^2/2g$. The exact value of C_v can be derived from Eqs. 3.17 and 3.18. In general terms for singular controls this is:

$$C_v = (H_1/h_1)^u = (1 + \alpha_1 v_1^2/2gh_1)^u \qquad (3.63)$$

where u equals the exponent of h_1 in the head-discharge equation for the studied range of discharges.

If the approach velocity v_1 is small, the velocity head $\alpha_1 v_1^2/2g$ is small with respect to H_1. Hence H_1 and h_1 are almost equal and the C_v value is just slightly more than unity. Because of the exponent 2 of v_1 in the velocity head, however, this head increases rapidly with increasing approach velocity.

3.6.2 C_v values for various control shapes

Substitution of $v_1 = Q/A_1$ into Eq. 3.63 gives:

$$C_v = \left(1 + \frac{\alpha_1 Q^2}{2gh_1 A_1^2}\right)^u \qquad (3.64)$$

The exact value of C_v for a given shape of the control section can be found by substituting into this equation the appropriate equation for Q and the related u value for the studied range of discharges.

<u>Rectangular control section:</u>

For a rectangular control section, u = 1.50, the discharge equals:

$$Q = C_d C_v \frac{2}{3} \left(\frac{2}{3}g\right)^{0.50} b_c h_1^{1.50} \qquad \text{(Eq. 3.18)}$$

Thus Eq. 3.64 becomes:

$$C_v = \left[1 + \frac{\alpha_1 C_d^2 C_v^2 \frac{4}{9} (\frac{2}{3} g)\, b_c^2\, h_1^3}{2gh_1 A_1^2}\right]^{1.50}$$

$$C_v = \left[1 + \frac{4\alpha_1 C_v^2}{27} \; \frac{C_d^2 (b_c h_1)^2}{A_1^2}\right]^{1.50} \qquad (3.65)$$

Bos (1976) plotted the C_v value of various control shapes versus the dimensionless area ratio $\sqrt{\alpha_1}\; C_d A^*/A_1$, where A* would be the imaginary area of the control section if the water depth were to equal h_1. For a rectangular control section (Figure 3.19):

$$A^* = b_c h_1 \qquad (3.66)$$

Substituting this A* into Eq. 3.65 and rewriting gives:

$$\sqrt{\alpha_1}\; C_d \frac{A^*}{A_1} = \sqrt{\frac{C_v^{2/3} - 1}{C_v^2} \; (\frac{27}{4})} = 2.60 \sqrt{\frac{C_v^{2/3} - 1}{C_v^2}} \qquad (3.67)$$

<u>Parabolic control section:</u>

For a parabolic control section, $A^* = \frac{4}{3} \sqrt{2f}\; h_1^{1.5}$, and the discharge equation is:

$$Q = C_d C_v \, (\frac{3}{4} gf)^{0.50}\; h_1^{\;2.0} \qquad \text{(Eq. 3.33)}$$

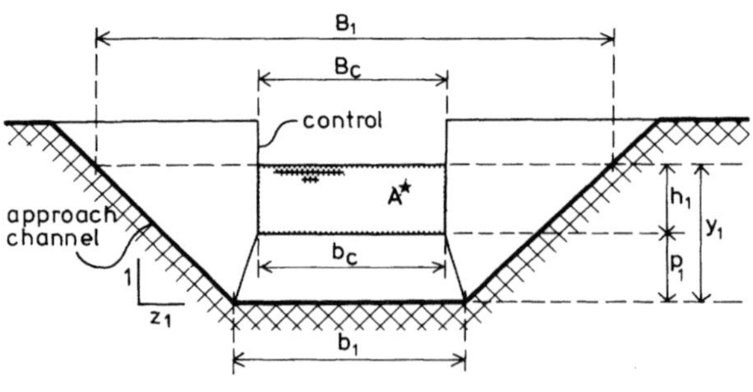

Figure 3.19. Section over gauging station and view of control

Substitution of this equation and u = 2.0 into Eq. 3.64 gives:

$$C_v = \left(1 + \frac{\alpha_1 C_d^2 C_v^2 \frac{3}{4} g f h_1^4}{2 g h_1 A_1^2}\right)^{2.0} \qquad (3.68)$$

or:

$$C_v = \left[1 + \frac{\alpha_1 C_d^2 C_v^2 (\frac{3}{4}) (\frac{9}{32}) A^{*2}}{2 A_1^2}\right]^{2.0} \qquad (3.69)$$

Rewriting yields:

$$\sqrt{\alpha_1}\, C_d A^*/A_1 = 3.08 \sqrt{\frac{C_v^{0.5} - 1}{C_v^2}} \qquad (3.70)$$

<u>Triangular control section:</u>

For a triangular control section, $A^* = h_1^2 \tan \frac{\theta}{2}$, and the discharge equation is:

$$Q = C_d C_v \frac{16}{25} (\frac{2}{5} g)^{0.50} \tan \frac{\theta}{2} h_1^{2.50} \qquad \text{(Eq. 3.21)}$$

Substitution of this equation and the related value u = 2.50 into Eq. 3.64 gives:

$$C_v = \left[1 + \frac{\alpha_1 C_d^2 C_v^2 (\frac{16}{25})^2 \frac{2}{5} g\, h_1 (h_1^2 \tan \frac{\theta}{2})^2}{2\, g\, h_1 A_1^2}\right]^{2.50} \qquad (3.71)$$

which, upon substitution of A*, reads:

$$C_v^{0.4} = 1 + \alpha_1 C_v^2 (\frac{16}{25})^2 (\frac{1}{5}) (C_d A^*/A_1)^2 \qquad (3.72)$$

Rewriting yields:

$$\sqrt{\alpha_1} C_d A^*/A_1 = 3.49 \sqrt{\frac{C_v^{0.4} - 1}{C_v^2}} \qquad (3.73)$$

3.6.3 Summary of C_v values

Equations 3.67, 3.70, and 3.73 give the relationship between the area ratio and the approach velocity coefficient. Substitution of various values of C_v into these equations yields the values of $\sqrt{\alpha_1} C_d A^*/A_1$ as given in Table 3.2 (Bos 1976, and Clemmens, Replogle, and Bos 1984).

Table 3.2. Table of values for plotting C_v versus $\sqrt{\alpha_1} C_d A^*/A_1$

Approach velocity coefficient	Area ratios $\sqrt{\alpha_1} C_d A^*/A_1$ for 3 control shapes		
	Rectangular	Parabolic	Triangular
C_v	$2.60\sqrt{\frac{C_v^{2/3}-1}{C_v^2}}$	$3.08\sqrt{\frac{C_v^{0.5}-1}{C_v^2}}$	$3.49\sqrt{\frac{C_v^{0.4}-1}{C_v^2}}$
1.00	0.000	0.000	0.000
1.05	0.450	0.461	0.470
1.10	0.606	0.619	0.625
1.15	0.707	0.721	0.728
1.20	0.779	0.793	0.800

Plotting the values of Table 3.2 in a graph yields Figure 3.20, which shows a remarkably close plot of C_v values as a function of the dimensionless area ratio. The explanation for this is that the imaginary area A* already expresses the control shape.

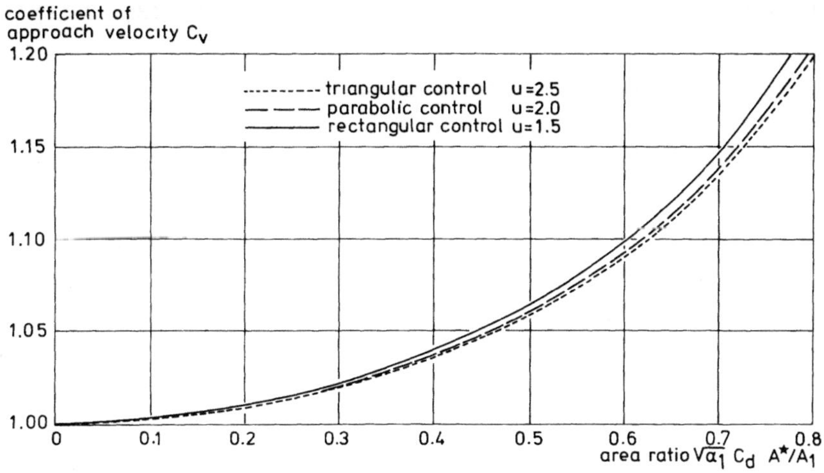

Figure 3.20. C_v values as a function of the area ratio $\sqrt{\alpha_1} C_d A^*/A_1$ (from Bos 1977)

3.7 ALTERNATIVE RATING BY ITERATION

3.7.1 Equations for ideal flow

The general equations for ideal flow at critical depth in the control section were given in Section 3.3.1. They are repeated here in a different setting. For ideal flow at constant rates through the weir of Figure 3.21, we can apply Eq. 3.5 to the gauging station. Hence, with $\alpha_1 = 1.0$:

$$H_1 = h_1 + Q_i^2/2gA_1^2 \tag{3.74}$$

where Q_i is the ideal flow rate.

Other equations that were derived for ideal flow read:

$$Q_i = A_c \left[2g\,(H_1 - y_c)\right]^{0.50} \qquad \text{(Eq. 3.12)}$$

in which accordingly:

$$y_c = H_1 - A_c/2B_c \qquad \text{(Eq. 3.13)}$$

The combination of these two equations gives:

$$Q_i = (g\, A_c^3/B_c)^{0.50} \qquad \text{(Eq. 3.58)}$$

This general equation is valid for all arbitrarily shaped control sections. As discussed in Section 3.5, the effect of energy losses and streamline curvature in the control section are accounted for by the introduction of a discharge coefficient, C_d:

$$Q = C_d Q_i \qquad \text{(Eq. 3.59)}$$

and for practical purposes:

$$C_d = 0.93 + 0.10\, H_1/L \qquad \text{(Eq. 3.61)}$$

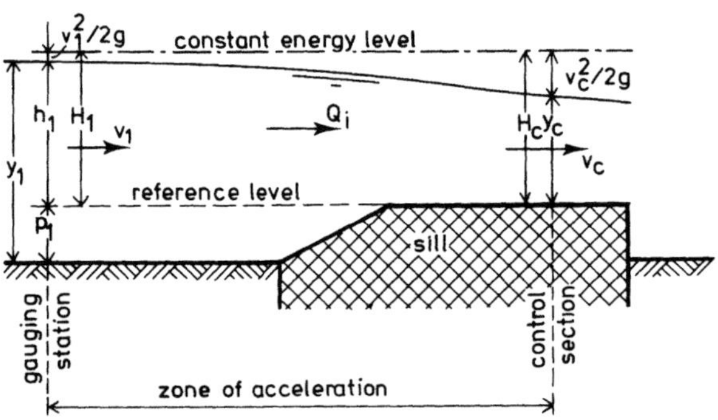

Figure 3.21. The energy level at gauging station and the control section for ideal flow.

3.7.2 The rating procedure

The combined use of the above equations is easy if simple equations exist for A_c and B_c in terms of y_c. For a trapezoidal control section these equations, for example, read:

$$A_c = y_c (b_c + z_c y_c) \tag{3.75}$$

and

$$B_c = b_c + 2 z_c y_c \tag{3.76}$$

The approach channel may have any shape. Most irrigation and drainage canals, however, are usually trapezoidal, or may be assumed to be so. Hence:

$$A_1 = y_1 (b_1 + z_1 y_1) \tag{3.77}$$

in which, as shown in Figure 3.21:

$$y_1 = p_1 + h_1 \tag{3.78}$$

Thus, for each combination of approach channel and control section shapes, the Eqs. 3.13, 3.58, and 3.74 have unknown y_c, Q_i, and h_1. If any one of these three is given, the other two can be solved by iteration.

The iteration procedure starts with determining the range of h_1 values for which the appropriate discharges need to be computed. Next, an initial guess is made for y_c in terms of h_1. As can be seen in Figure 3.12, y_c ranges from 0.67 H_1 for a rectangular control section ($z_c = 0$) to 0.80 H_1 for a triangular control section ($b_c = 0$). If the velocity head, $v_1^2/2g$, is neglected, an initial guess for y_c could be:

$$y_c = 0.70\ h_1 \tag{3.79}$$

No closer guess of y_c is needed for different shapes of the control section, since the iteration process converges rapidly. Once y_c has been guessed, values of A_c, B_c, and Q_i can be computed, followed by those of H_1 and y_c (from computed Q_i value). If the new y_c value equals the input y_c value then the computed Q_i is the correct flow rate for an ideal fluid matching the set h_1 value. Usually, however, the new

y_c value will not equal the input value, and a new series of calculations will have to be made until the y_c values match. This iteration loop is illustrated in Figure 3.22.

The actual flow rate, Q, that matches the set h_1 value is found by

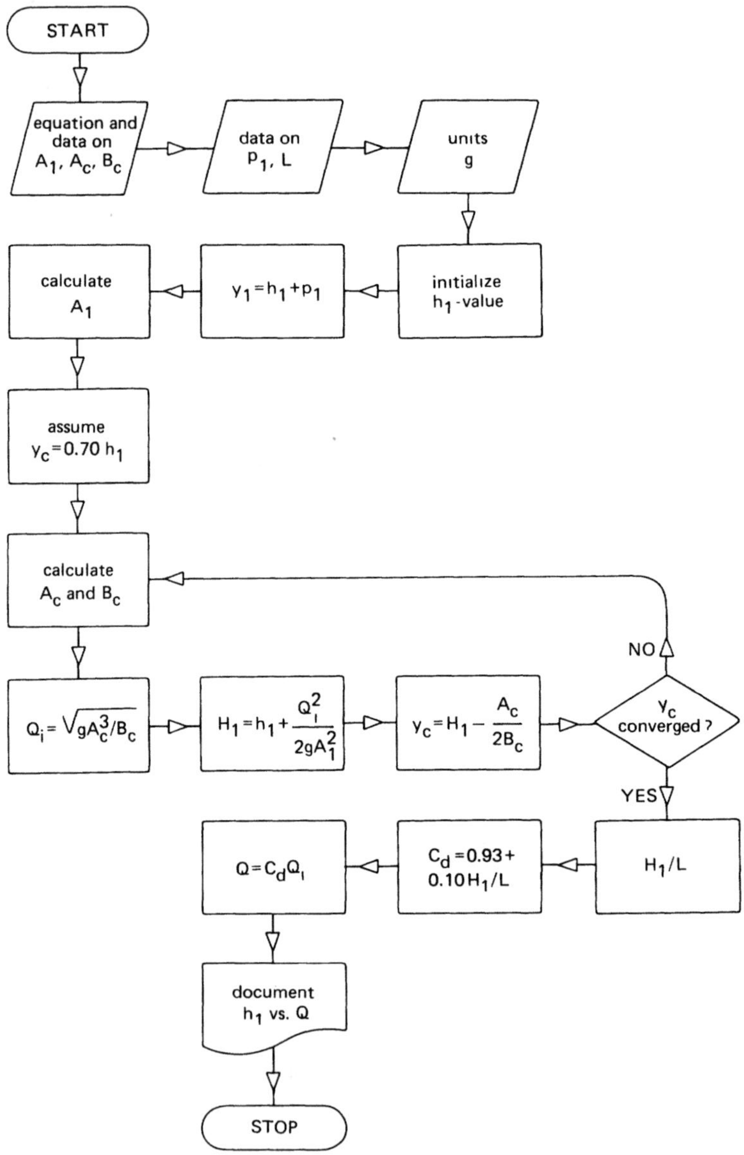

Figure 3.22. Flow chart for the calculation of head-discharge (h_1-Q)

using the last H_1 value for the calculation of H_1/L and the related C_d value from Eq. 3.64. Finally, Eq. 3.59 is used to calculate the actual flow Q.

This method has the advantage of not requiring an estimate of C_v, since both H_1 and h_1 are used in the computation, and the energy heads are balanced. A further advantage is that because the method starts with h_1 rather than H_1, a programmable calculator can be used for the direct development of head-discharge relationships.

4
Required head loss over the structure

4.1 INTRODUCTION

As stated when Eqs. 3.8 and 3.12 were being introduced, the water depth at the control section of a structure must be critical to give a unique relation between the upstream sill-referenced head and the flow rate in the structure. If the downstream sill-referenced head is sufficiently low with respect to the upstream sill-referenced head, modular flow will occur. With high downstream heads, the flow at the control section cannot become critical and the downstream sill-referenced head influences the upstream sill-referenced head. If the submergence ratio, H_2/H_1, is high, the flow is non-modular.
That ratio of H_2/H_1 at which flow at the control section can just become critical is called the modular limit, ML, for the corresponding rate of flow.

The loss of head, ΔH, is caused by friction and turbulence losses in the structure and the adjoining channel reaches from the upstream to the downstream head measurement sections. This loss of energy head can be related to the upstream head in terms of the submergence ratio by:

$$(H_1 - H_2)/H_1 = 1 - H_2/H_1 \qquad (4.1)$$

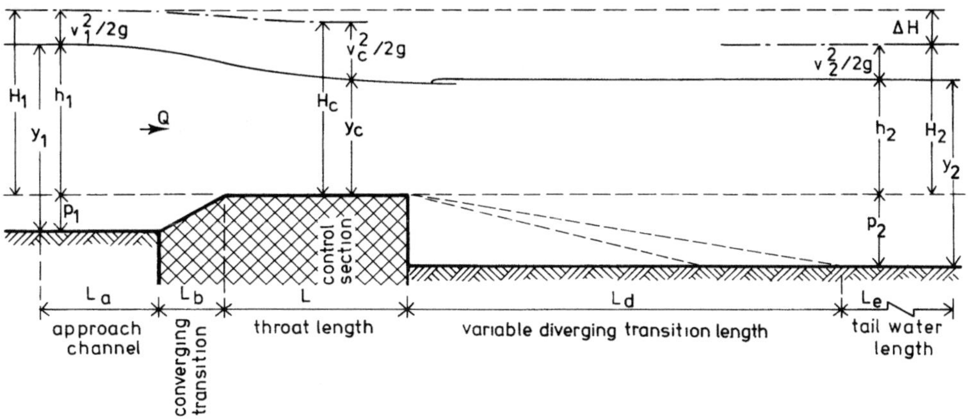

Figure 4.1. Illustration of terminology.

To design a structure with modular flow in a given channel, the following variables must be selected:

(i) The elevation of the weir sill or flume throat;

(ii) The shape and related width and depth of the control section;

(iii) The expansion ratio and the length of the diverging transition.

These design variables greatly influence the head loss requirement for modular flow, making the process of matching the structure to the (depth-discharge curve of the) channel an iterative one. In this context, it should be noted that the channel does not have a stable depth-discharge curve. If this were the case, no discharge measuring structure would be required. The design procedure, however, means choosing a measuring structure with an optimal stage-discharge relationship.

For modular flow, the head-discharge equations of Figure 3.12 can be used. For non-modular flow, the actual discharge is less than that calculated by these equations. To correct for this reduction in flow rate, a 'drowned flow reduction factor', f, can be used so that:

$$Q_{non\text{-}modular} = fQ_{modular} \tag{4.2}$$

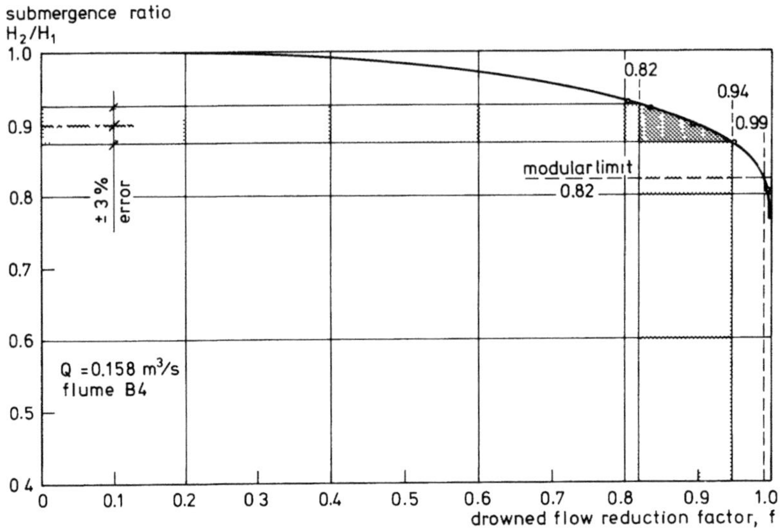

Figure 4.2. Example of the relationship between drowned flow reduction factor and submergence ratio that varies with a rising downstream water level.

Figure 4.2 shows a typical curve giving f values as a function of the submergence ratio for a trapezoidal flume (B4) as tested by Bos and Reinink (1981) at a constant flow rate of Q = 0.158 m^3/s. The higher the value of the modular limit, the more the curve will move into the top right corner of the figure. In this thesis, the modular limit is numerically defined as the value of the submergence ratio H_2/H_1 at which the actual discharge deviates by 1% from the modular discharge (f = 100/101 ≈ 0.99).

Flumes and weirs with non-modular flow are not recommended for measuring discharge. The reasons are:

1. The drowned flow reduction factor is unknown unless the given combination of control section, diverging transition and tailwater channel is tested in a hydraulic laboratory for the complete range of relevant flow rates and downstream heads;
2. The value of f for a given structure is not a function of H_2/H_1 only, but also of the rate of flow, which has yet to be measured;

3. Errors in determining both H_1 and H_2 propagate in H_2/H_1. In the example of Figure 4.2, an error of ± 3% in the submergence ratio causes the f value to vary from 0.82 to 0.94. In practice, the error in H_2/H_1 often exceeds 3%. Especially in cases where the modular limit is high, the f value cannot be determined with sufficient accuracy;
4. Non-modular flow requires the measurement of two sill-referenced heads (h_1 and h_2). Processing these heads into energy heads and combining them with a drowned flow reduction factor - a function of the Q to be measured - can be an expensive trial-and-error procedure.

4.2 THEORY

The fundamental condition for flow at the modular limit is that the available loss of head between the channel cross-sections where the upstream head, H_1, and the downstream head, H_2, are to be determined, is just sufficient to satisfy the requirement for critical flow to occur at the control section. This situation will be analyzed by dividing this minimum loss of energy head, $H_1 - H_2$, into three parts (Bos 1976):

(1) The energy head loss, $H_1 - H_c$, between the upstream head measurement section (gauging station) and the control section in the flume throat (Section 4.2.1);

(2) The energy losses, ΔH_f, due to friction between the control section and the downstream head measurement section (Section 4.2.2);

(3) The losses, ΔH_d, due to turbulence in the diverging transition (Section 4.2.3).

Figure 4.3 indicates the lengths of those parts of the structure for which these three energy losses are to be calculated.

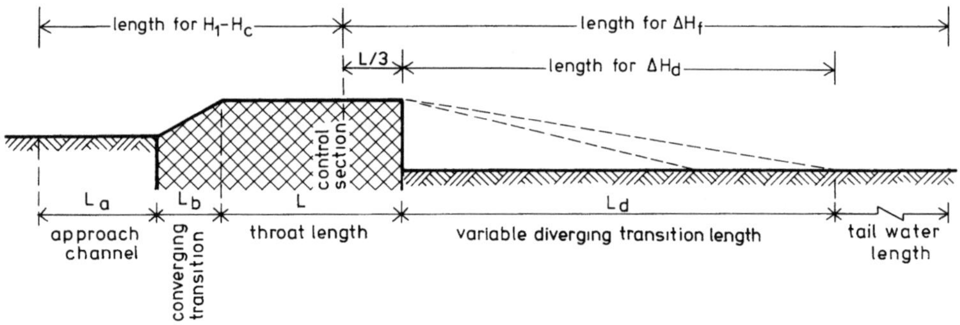

Figure 4.3. Lengths of structure parts for which $H_1 - H_c$, ΔH_f, and ΔH_d are to be calculated.

4.2.1 Energy losses upstream of the control section

The head-discharge relationship for a rectangular, parabolic, or triangular control section, and for parts of all other control section shapes, can be written in the exponentional form:

$$Q = C_d K H_1^{\ u} = C_d \ C_v K h_1^{\ u} \tag{4.3}$$

where K is a dimensional coefficient, a constant for a given weir or flume. The value of the exponent, u, can be calculated from:

$$u = \frac{h_1}{Q} \cdot \frac{dQ}{dh_1} \tag{4.4}$$

For the application of this equation, a rating curve of the considered control section should be available on linear paper (Figure 4.4). During the design process, however, such a curve is not available. The exponent, u, can then be approximated from:

$$u = \frac{\log Q_b - \log Q_a}{\text{loh } h_{1,b} - \log h_{1,a}} \tag{4.5}$$

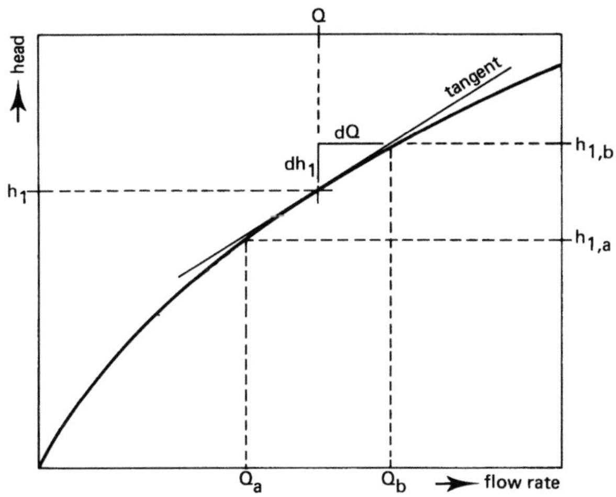

Figure 4.4. Illustration of terms in Eqs. 4.4 and 4.5

For the application of Eq. 4.5, the flow rates Q_a and Q_b must be calculated from the valid head-discharge equation (see Chapter 3). The choice of the appropriate values for the discharge coefficient, C_d, was discussed in Chapter 3. As stated when C_d was being introduced in Eq. 3.17, its value follows from the need to correct for:

(i) Energy losses between the gauging station and the control section;

(ii) The effect of curvature of streamlines in the control section;

(iii) The non-uniformity of the velocity distribution in both sections.

For heads that are low with respect to the throat length, the influence of streamline curvature and of the non-uniformity of the velocity distribution is negligible with respect to the energy losses (Ackers and Harrison 1963; Replogle 1975; Bos 1976; Bos and Reinink 1981). Consequently, it can be assumed that C_d only expresses the energy losses between the gauging station and the control section. Acting on this assumption and replacing H_1 by H_c in Eq. 4.5 results in:

$$Q = KH_c^{\,u} \tag{4.6}$$

Combining Eqs. 4.5 and 4.6 gives:

$$H_c^{\,u} = C_d H_1^{\,u} \tag{4.7}$$

which can also be written as (Bos 1976):

$$H_1 - H_c = H_1 \left(1 - C_d^{\,1/u}\right) \tag{4.8}$$

The right-hand member of this equation approximates the loss of hydraulic energy between the gauging and control sections. This equation, however, is only valid if the influence of streamline curvature at the control section on the C_d value is insignificant. This limitation is investigated in Section 4.3.3.

4.2.2 Friction losses downstream of the control section

Although flow is non-uniform in the diverging transition, the energy losses due to friction are estimated by applying the Manning equation to the three reaches shown in Figure 4.3:

(1) Reach of the flume throat downstream of the control section; the length of this reach is held at L/3;

(2) Length of the reach of the actual diverging transition of bottom and side walls, L_d;

(3) Length of a canal reach from the end of the transition to the measurement section of the downstream sill-referenced head ($L_e \simeq 5\, y_2$).

So that:

$$\Delta h_{throat} = \frac{1}{3} L \left(\frac{n\, Q}{A_c\, R_c^{\,2/3}}\right)^2 \tag{4.9}$$

$$\Delta H_{trans} = L_d \left(\frac{n\, Q}{A_d\, R_d^{\,2/3}}\right)^2 \tag{4.10}$$

and:

$$\Delta H_{canal} = L_c \left(\frac{n\, Q}{A_2\, R_2^{\,2/3}}\right)^2 \tag{4.11}$$

In the calculation of ΔH_{trans} and average area of flow, $A_d = (A_c + A_2)/2$ can be used. The n value in each of the equations depends on the construction material of the related reach of the structure and canal.
As stated in Section 3.5.2, the structure is assumed to be smoothly. finished.
The total energy losses due to friction, ΔH_f, between the control section and the section where h_2 is measured then equals the sum of the losses over the three reaches:

$$\Delta H_f = \Delta H_{throat} + \Delta H_{trans} + \Delta H_{canal} \qquad (4.12)$$

In contrast to the dimensions of the area of flow in the approach channel and the control, the dimensions of the downstream area of flow depend on the unknown value of H_2. The calculation of the modular limit therefore requires the solution by iteration of an implicit function of the downstream head (Section 4.2.5).

4.2.3 Losses due to turbulence in the zone of deceleration

In the diverging transition, part of the kinetic energy is converted into potential energy. The remainder is lost in turbulence. With flow at the modular limit, losses due to turbulence in the hydraulic jump are low (Peterka 1958) so the simple classical expression of Borda for energy losses in an expansion of a closed conduit can be used:

$$\Delta H_d = H_c - H_2 - \Delta H_f = \xi \frac{(v_c - v_2)^2}{2g} \qquad (4.13)$$

in which:

ξ = the energy loss coefficient, being a function of the expansion ratio of the diverging transition;

$v_c - v_2$ = decrease in average flow velocity between the control section and the downstream head measurement section.

Here again, v_2 depends on the unknown downstream head, H_2, so that the solution of Eq. 4.13 is part of the iteration process (Section 4.2.5).

International literature contains few data that allow the measured total energy loss over flumes to be broken down into the above three parts and permit ξ-values to be calculated. Blau (1960), Engel (1934), Inglis (1929), and Fane (1927), however, published sufficient data on the geometry of structures and channels to allow the total head loss, ΔH, to be broken down into a friction part and a turbulence part. The calculated ξ values that were obtained from this literature are shown in Figure 4.7. They correspond with the ξ values for the B and C series of the experiments conducted by Bos and Reinink (1981). The results of these experiments are discussed in Section 4.3.4.

4.2.4 Total energy loss requirement

The total energy loss over a flume or weir at the modular limit can be estimated by adding the three component parts as discussed in the preceding sections:

$$H_1 - H_2 = H_1 (1 - C_d^{1/u}) + \Delta H_f + \xi (v_c - v_2)^2/2g \qquad (4.14)$$

For the considered rate of flow through the structure, this equation gives the minimum loss of energy head required for modular flow. That part of the above equation which expresses the sum of the energy losses due to friction, $H_1 (1 - C_d^{1/u}) + \Delta H_f$, becomes a large percentage of the total energy loss, $H_1 - H_2$, when diverging transitions are long (high ΔH_f values). This is mainly because the relatively high flow velocities in the downstream transition are maintained over a greater length. On the other hand, very gradual downstream transitions have a favourable energy conversion (low ξ value). As a result, very gradual transitions may, as a whole, lose more energy than more rapid but shorter transitions. Since, in addition, the construction cost of a very gradual transition is higher than that of a shorter one, there are good arguments in favour of limiting the ratio of expansion to about 6-to-1.

Rather sudden expansion ratios like 1 to 1 or 2 to 1 are not effective because the high velocity jet leaving the throat cannot suddenly change direction to follow the boundaries of the transition. In the resulting flow separation zones, turbulence converts kinetic energy into head and noise.

If for any reason the channel cannot accommodate a fully developed gradual transition of 6-to-1 it is recommended that the transition be truncated to $L_d = H_{1max}$ rather than to use a more sudden expansion ratio (see Figure 2.5). The end of the truncated expansion should not be rounded, since it guides the water into the channel boundary; a rounded end causes additional energy losses and possible erosion.

The modular limit of a weir or flume can be found by dividing both sides of Eq. 4.14 by H_1, giving:

$$(H_2/H_1)_{at\ ML} = C_d^{1/u} - \Delta H_f/H_1 - \xi\,(v_c - v_2)^2/2gH_1 \qquad (4.15)$$

Equation 4.15 is a general expression for the modular limit of any long-throated flume, and is also valid for the hydraulically similar broad-crested weir.

4.2.5 Procedure to estimate the modular limit

Modularity of a discharge measurement structure in a given channel implies that the required head loss for modular flow must be less than the available head loss. The required head loss, however, can only be estimated after the type and dimensions of the structure have been chosen.

As will be explained in Chapter 5, this choice is also governed by other considerations, such as range of flows to be measured, and accuracy. An optimal solution to the design problem can only be found in an iterative design procedure (Section 5.8). Part of this procedure is estimating the modular limit for Q_{max} and Q_{min} of the tentative design under consideration in the relevant design loop.

In the following procedure to estimate the modular limit, it is assumed that the relationship between Q, h_1, and C_d is known. If this relationship is unknown, it can be derived as explained in Chapter 3. To estimate the modular limit of a weir or flume in a channel of given cross-section, both sides of Eq. 4.15 must be equalized as follows:

1. Determine the cross-sectional area of flow at the station where h_1 is measured, and calculate the average velocity, v_1;
2. Calculate $H_1 = h_1 + v_1^2/2g$;
3. For the given flow rate and related head, note down the C_d value;
4. Determine the exponent u;
   For a rectangular (u = 1.5), parabolic (u = 2.0), or triangular control section (u = 2.5), the power u is known from the head-discharge equation. For all other singular or composite control shapes, use Eq. 4.4 or 4.5;
5. Calculate $C_d^{1/u}$;
6. Use Section 3.3 and Figure 3.12 to find y_c at the control section. Note that y_c is a function of H_1 and of the throat size and shape;
7. Determine the cross-sectional area of flow at the control section with the water depth, y_c, and calculate the average velocity, v_c;
8. Use Figure 4.7 to find an ξ value as a function of the angle of expansion;
9. Estimate the value of h_2 that is expected to suit the modular limit and calculate A_2 and the average velocity v_2;
10. Calculate $\xi(v_c - v_2)^2/2gH_1$;
11. Determine $\Delta H_f = \Delta H_{throat} + \Delta H_{trans} + \Delta H_{canal}$ by applying the Manning equation with the appropriate value of n to $\frac{1}{3}L$ of the throat, to the transition length, and to the canal up to the h_2 measurement section (see Section 4.2.2);
12. Calculate $\Delta H_f/H_1$;
13. Calculate $H_2 = h_2 + v_2^2/2g$;
14. Calculate H_2/H_1;
15. Substitute the values (5), (10), (12), and (14) into Eq. 4.15;
16. If Eq. 4.15 does not match, repeat steps (9) through (15).

Once some experience has been acquired, Eq. 4.15 can be solved with two or three iterations. Since the modular limit varies with the upstream head, it is advisable to estimate the modular limit at both minimum and maximum anticipated flow rates and to check if sufficient head loss ($H_1 - H_2$) is available in both cases (see also Section 5.3.1).

4.3 HYDRAULIC LABORATORY TESTS

To verify the theory and related procedure of estimating the modular limit (Section 4.2), model tests were conducted in the hydraulic laboratory of the University of Agriculture at Wageningen (Bos and Reinink 1981).

4.3.1 Description of model tests

In a concrete-lined channel section with bottom $b_1 = b_2 = 0.50$ m and a side slope of 1-to-1, a trapezoidal long-throated flume was constructed (Figure 4.5). The flume throat was made of polyvinyl chloride (PVC) with a sill height $p_1 = p_2 = 0.15$ m, length L = 0.60 m, width $b_c = 0.20$ m, and a side slope $z_c = 1.0$. This model was first used to study the effect on the C_d value of a 1-to-2 slope of the converging transition versus the commonly used 1-to-3 slope. The result of these A series of tests are given in Section 3.5.3. In a further series of tests, the B series, the bottom of the throat was again placed 0.15 m above the bottom of the approach channel bottom so that $p_1 = p_2 = 0.15$ m (see Figure 4.5).
In each test series, the flume was successively fitted with six diverging transitions as shown in Table 4.1. During Test B4, some difficulties arose with instability of the tailwater level, h_2. To solve this problem, a piezometer was placed to measure h_2, and for Test B5 the flume was moved 1.40 m upstream (see Figure 4.5).

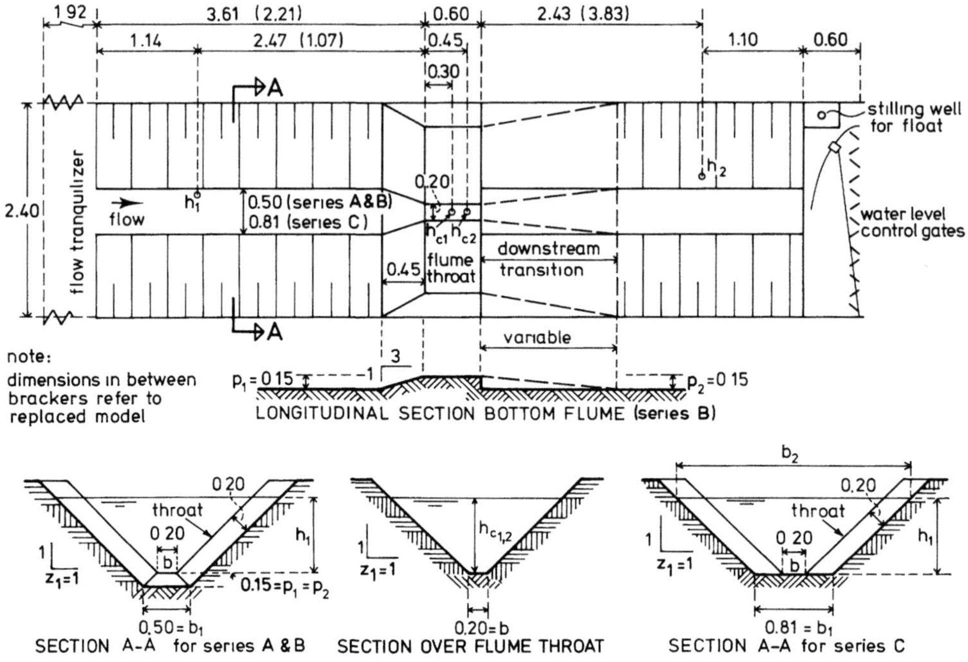

Figure 4.5. Dimensions of model, in metres (Bos and Reinink 1981)

For the C series, the bottoms of the approach and tailwater channels were raised by 0.15 m so that $p_1 = p_2 = 0$. In this series, the flat-bottomed flumes thus needed no transition on the bottom. The ratios of divergence of the side wall transitions are the same as those in the B series.

Piezometers were also installed in the bottom of the throat to allow a direct derivation of the local energy head, h_c.

Table 4.1. Diverging transitions for tests of B series

	Test number					
	B1	B2	B3	B4	B5	B6
Bottom slope	0-to-1	1-to-1	2-to-1	4-to-1	6-to-1	10-to-1
length of diverging transition, L_d	0	0.15	0.30	0.60	0.90	1.50
Side wall slope	0-to-1	0.7-to-1	1.43-to-1	2.9-to-1	4.3-to-1	7-to-1

4.3.2 Measuring the modular limit

To evaluate the effect of submergence at various flow rates on the related upstream sill-referenced head of the tested flumes, the Q-H_1 relationship around the considered flow rate is written in the general form:

$$Q = K_* H_1^{u} \qquad (4.16)$$

where:

u = the dimensionless power of H_1, the value of which depends on the shape of the cross-section of the flume throat and the range of considered flow rates (see Section 4.2.1);

K_* = a dimensional coefficient whose value depends on the size and shape of the flume (its dimension depends on u).

Equation 4.16 implies that the log Q versus log H_1 curve is straight. This is true for all rectangular, parabolic, and triangular control sections, and it is at least a good approximation for all other control sections if the studied range of Q is limited.
For each of a number of fixed flow rates, calculations were made of the head $H_{1(101)}$ at which the flow rate, as calculated by the head-discharge equation for modular flow, deviated by 1% from the actual flow rate in the test run:

$$1.01 = Q_{(101)}/Q_{(100)} = K_* H_{1(101)}^{u}/K_* H_{1(100)}^{u} \qquad (4.17)$$

$$H_{1(101)}^{u} = 1.01\, H_{1(100)}^{u} \qquad (4.18)$$

$$\log H_{1(101)} = \frac{1}{u}\left[\log 1.01 \times H_{1(100)}^{u}\right] \qquad (4.19)$$

After $H_{1(101)}$ had been calculated with this equation, the related limiting value of the upstream sill-referenced head was found from:

$$h_{1(101)} = H_{1(101)} - v_1^{2}/2g \qquad (4.20)$$

Table 4.2. Modular limits and energy losses as a function of flow rate in twelve trapezoidal flumes (Bos and Reinink 1981)

Flume	Data	Discharge, in cubic metres per second			
		0.010	0.030	0.090	0.160
B1	h_2/h_1	0.7564	0.8288	0.7885	0.7075
	H_2/H_1	0.7615	0.8353	0.8006	0.7293
	ΔH	0.0203	0.0253	0.0549	0.0963
B2	h_2/h_1	0.7758	0.8178	0.7987	0.7020
	H_2/H_1	0.7812	0.8242	0.8104	0.7241
	ΔH	0.0184	0.0269	0.0522	0.0981
B3	h_2/h_1	0.7927	0.8292	0.8023	0.7183
	H_2/H_1	0.7980	0.8354	0.8138	0.7391
	ΔH	0.0171	0.0253	0.0513	0.0927
B4	h_2/h_1	0.8176	0.8548	0.8136	0.7717
	H_2/H_1	0.8230	0.8607	0.8246	0.7901
	ΔH	0.0148	0.0214	0.0483	0.0745
B5	h_2/h_1	0.8145	0.8567	0.8503	0.8216
	H_2/H_1	0.8218	0.8629	0.8599	0.8355
	ΔH	0.0152	0.0211	0.0385	0.0584
B6	h_2/h_1	0.8361	0.8757	0.8705	0.8416
	H_2/H_1	0.8414	0.8874	0.8796	0.8543
	ΔH	0.0135	0.0173	0.0331	0.0519
C1	h_2/h_1	0.7730	0.8087	0.8906	0.8309
	H_2/H_1	0.7873	0.8232	0.9030	0.8519
	ΔH	0.0184	0.0275	0.0268	0.0531
C2	h_2/h_1	0.7860	0.8222	0.8855	0.8333
	H_2/H_1	0.7991	0.8367	0.8983	0.8537
	ΔH	0.0173	0.0254	0.0281	0.0525
C3	h_2/h_1	0.7937	0.8390	0.8739	0.8301
	H_2/H_1	0.8068	0.8527	0.8880	0.8509
	ΔH	0.0166	0.0228	0.0309	0.0533
C4	h_2/h_1	0.8113	0.8497	0.8689	0.8402
	H_2/H_1	0.8231	0.8620	0.8850	0.8602
	ΔH	0.0152	0.0231	0.0321	0.0501
C5	h_2/h_1	0.8271	0.8702	0.8528	0.8654
	H_2/H_1	0.8388	0.8812	0.8690	0.8824
	ΔH	0.0138	0.0183	0.0360	0.0421
C6	h_2/h_1	0.8436	0.8901	0.8642	0.8827
	H_2/H_1	0.8541	0.9002	0.8790	0.8978
	ΔH	0.0124	0.0154	0.0334	0.0366

Note: With flumes B1, B2, and B3, the h_2 value was measured at 1.90 m downstream of the throat, for flume B4 at 2.43 m, and for the other flumes at 3.18 m (see Figure 4.5).

This value of $h_{1(101)}$ is the minimum value of h_1 that is required to preserve modularity at the prevailing head, h_2.

For each of the twelve tested flumes in the B and C series, the tests were run at flow rates as close as possible to 0.10 m^3/s, 0.030 m^3/s, 0.090 m^3/s, and 0.160 m^3/s respectively. The downstream water level was raised stepwise. When the flow pattern had stabilized, the actual flow rate was read on the flow meter, and all heads were measured with point gauges. This process yielded the limiting downstream water level, h_2, which caused the calculated value of $h_{1(101)}$, so that the limiting submergence ratio h_2/h_1 could be calculated.
Table 4.2 lists the modular limits for each combination of flume and discharge in terms of both measured sill-referenced heads, h_2/h_1, and the calculated energy heads, H_2/H_1.

4.3.3 Head loss $H_1 - H_c$

Equation 4.8 was derived to estimate the loss of hydraulic energy between the gauging station and the control section, subject to the assumption of straight and parallel streamlines in the control section. To evaluate the validity of this assumption, laboratory data were used to calculate $H_1 = h_1 + v_1^2/2g$, $H_c = y_c + v_c^2/2g$, C_d, and u for fixed flow rates. The resulting values allowed both sides of Eq. 4.8 to be calculated separately for twelve long-throated flumes with trapezoidal cross-section (B and C in Section 4.3.1). Of these twelve flumes, six had a bottom hump of $p_1 = p_2 = 0.15$ m; the others had a flat bottom ($p_1 = p_2 = 0$). The diverging transition varied as shown in Table 4.1. Figure 4.6 shows that the energy loss upstream of the control section of the B and C flumes can be estimated by the right-hand member of Eq. 4.8 if this loss is less than about 5 mm. For the twelve tested flumes, this loss of 5 mm occurred at flow rates that caused an H_1/L value of about 0.5.

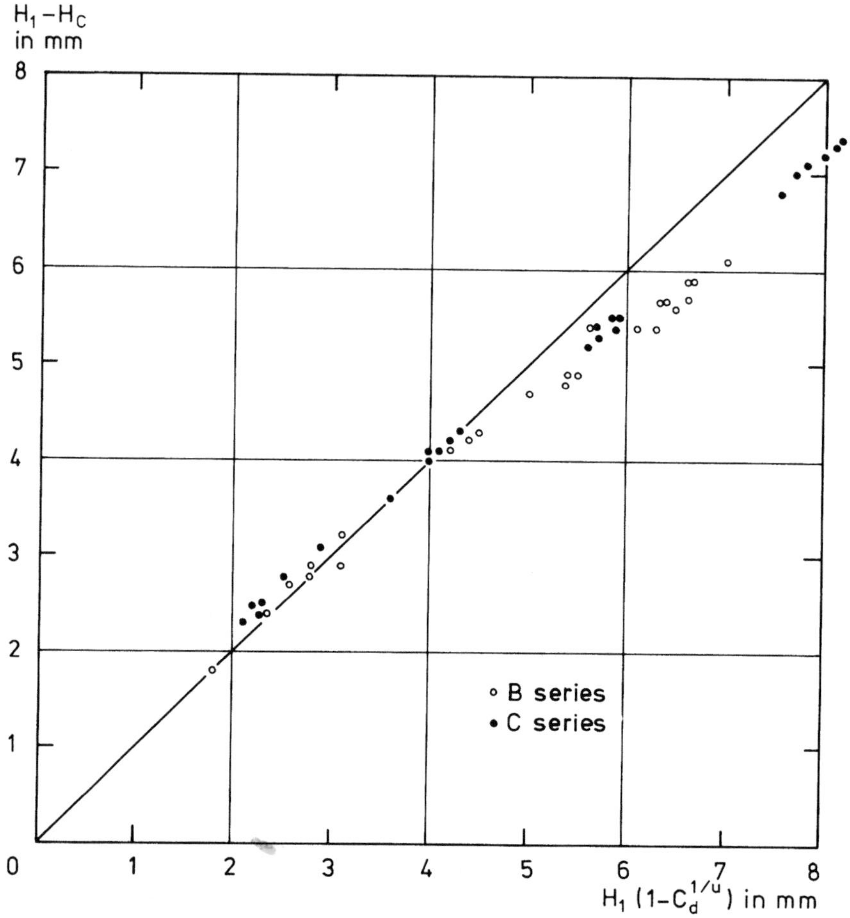

Figure 4.6. Head loss due to friction between the gauging and control sections (Bos and Reinink 1981)

Hence Eq. 4.8 can be used to estimate $H_1 - H_c$ if the H_1/L ratio is less than 0.5. For higher values of H_1/L, streamlines at the control section are significantly curved, so that the concept of minimum specific energy, $H_c = y_c + v_c^2/2g$, is no longer valid and Eq. 4.8 does not apply (see Section 4.4).

4.3.4 Energy losses downstream of the control section

As discussed in Section 4.2, the energy head loss, ΔH, for flow at the modular limit can be divided into three parts: $H_1 - H_c$, ΔH_f, and ΔH_d. Table 4.3 lists the measured values of ΔH, the calculated values of $H_1 - H_c$ (by Eq. 4.8) and those of ΔH_f (by Eq. 4.12), and the remaining energy head loss ΔH_d. Because of the approximate nature of the assumptions that underlie the application of both Eqs. 4.8 and 4.12, the number of significant figures given for $H_1 - H_c$ and ΔH_f does not imply a corresponding accuracy. They serve merely to allow data to be compared and to calculate ΔH_d.

As can be seen in Table 4.3, the total energy loss due to friction, $(H_1 - H_c) + \Delta H_f$, expressed as a percentage of the total energy loss over the flume, ΔH, increases with increasing length of the diverging transition. For a flat-bottomed flume with a 10-to-1 transition (C6), this percentage is already greater than 35. Because of the relatively high flow velocities in the diverging transition in a flat-bottomed flume (C series), the percentage of energy loss due to friction is larger than in a flume with an elevated throat (B series).
From the calculated ΔH_d values of Table 4.3, and the related average velocities, v_c and v_2, the energy loss coefficient, ξ, for the tested flumes was calculated with Eq. 4.13. As shown in Figure 4.7, these values correspond with the ξ values for the structures tested by Fa (1927), Inglis (1929), Engel (1934), and Blau (1960).

Table 4.3. Energy losses over flumes flowing at their modular limit, in centimetres (Bos and Reinink 1981)

	Tested flume	Discharge, in cubic metres per second		Tested flume	Discharge, in cubic metres per second	
		0.010	0.030		0.010	0.030
$\Delta H = H_1 - H_2$	B1	2.03	2.53	C1	1.84	2.75
$H_1 - H_c$		0.29	0.47		0.24	0.42
ΔH_f		0.07	0.07		0.19	0.18
ΔH_d		1.67	1.99		1.41	2.15
ΔH	B2	1.84	2.69	C2	1.73	2.54
$H_1 - H_c$		0.27	0.43		0.25	0.43
ΔH_f		0.07	0.07		0.19	0.18
ΔH_d		1.50	2.19		1.29	1.93
ΔH	B3	1.71	2.53	C3	1.66	2.28
$H_1 - H_c$		0.32	0.54		0.23	0.36
ΔH_f		0.07	0.07		0.19	0.18
ΔH_d		1.32	1.92		1.24	1.74
ΔH	B4	1.48	2.14	C4	1.52	2.31
$H_1 - H_c$		0.18	0.46		0.31	0.40
ΔH_f		0.08	0.08		0.18	0.18
ΔH_d		1.22	1.60		1.03	1.73
ΔH	B5	1.52	2.11	C5	1.38	1.83
$H_1 - H_c$		0.29	0.41		0.25	0.41
ΔH_f		0.08	0.08		0.18	0.17
ΔH_d		1.15	1.62		0.95	1.25
ΔH	B6	1.35	1.73	C6	1.24	1.54
$H_1 - H_c$		0.24	0.28		0.27	0.41
ΔH_f		0.08	0.09		0.17	0.16
ΔH_d		1.03	1.36		0.80	0.97

To estimate the energy loss connected with conversion of kinetic energy into potential energy in the diverging transition of a weir or flume, the use of the envelope of Figure 4.7 is recommended. This conservative ξ value is then substituted into Eq. 4.13 to estimate ΔH_d (see Section 4.2.5).

Because the increasing friction losses in a more gradual expansion are only partly compensated for by a lower ξ value, the required energy loss for modular flow, ΔH, decreases little if the expansion ratio exceeds a certain value. Weighing the construction cost of a more gradual expansion against the lower requirement of ΔH yields a practical upper limit of the expansion ratio, EM, of 6-to-1.

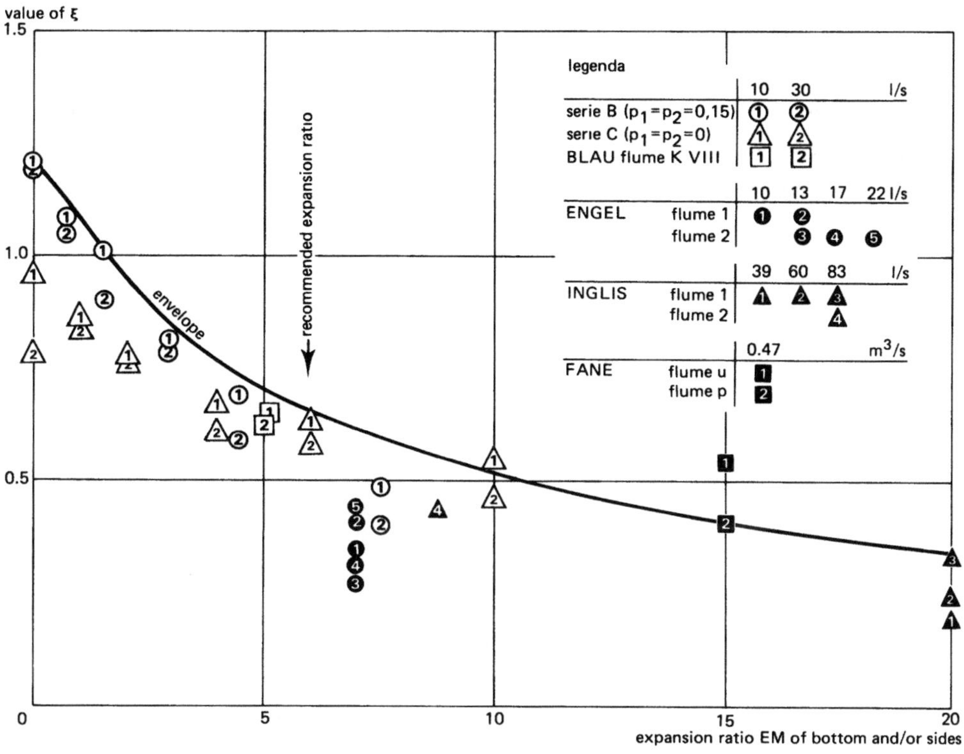

Figure 4.7. Values of ξ as a function of the expansion ratio of downstream transition (Bos and Reinink 1981)

4.4 EXPERIMENTAL VERIFICATION OF MODULAR LIMIT ESTIMATE WITH RECENT TESTS

To evaluate the presented method of estimating the modular limit of a long-throated flume, data from recent literature were available for use. From the actual dimensions of laboratory models (Boiten 1983; Dodge 1982; ISO 1983; Smith and Liang 1969), and of a large prototype (Replogle et al. 1983) the modular limit was estimated. These results were plotted against the modular limits as measured by the above authors (Figure 4.8). For three combinations of structure and flow rate, the method overestimated the modular limit. All three were weirs with bottom contraction only, placed in a rectangular laboratory flume and flowing at $H_1/L \approx 0.5$.

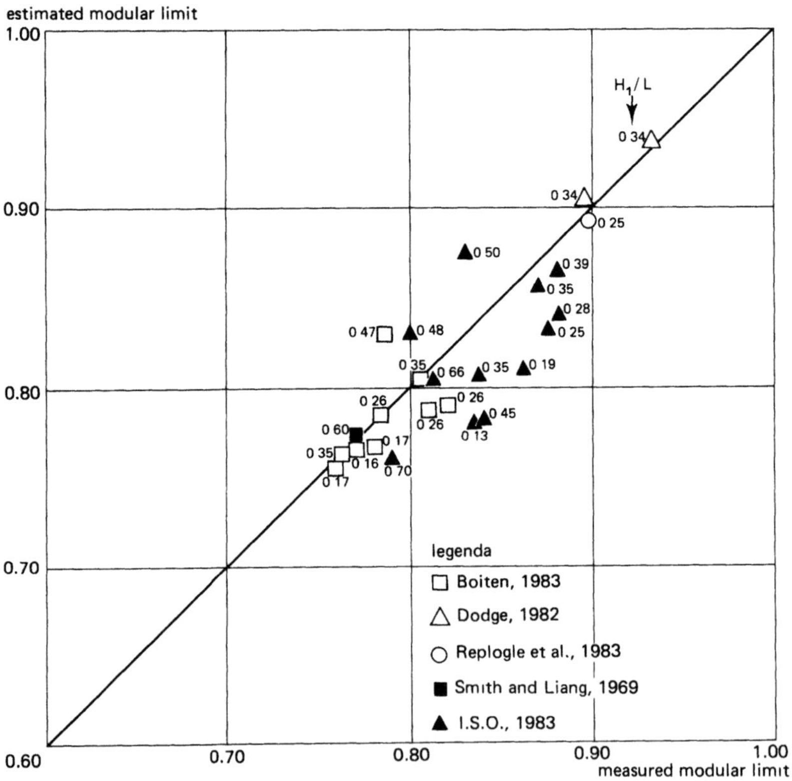

Figure 4.8. Measured versus computed modular limits

Figure 4.8 shows that the method of estimating the modular limit is a sufficiently accurate approximation for the design of long-throated flumes and the hydraulically related broad-crested weirs. As discussed in Section 4.2.1 and 4.3.3, the related theory is valid, provided that streamline curvature at the control section is negligible; hence if $H_1/L < 0.50$. With higher H_1/L ratios, the streamlines at the control section become increasingly curved, which has a negative effect on the modular limit.
As a conservative rule, it is assumed that if the H_1/L ratio increases from 0.5 to 1.0, the modular limit decreases linearly from the value calculated to a value of y_c/H_1, where y_c is the critical depth in the control section. This rule cannot be supported by the results of systematic experiments. It is supported, however, by observations made by Bos 1976, Bos and Reinink 1981, Boiten 1983, and ISO 1983.

4.5 VISUAL DETECTION OF MODULAR LIMIT

If the tailwater level is sufficiently low for a stable hydraulic jump to occur downstream of the control section, this jump is clearly visible (see top row of Photo 4.1). Because of the hydraulic jump, flow must be critical at the control section and thus be modular.

With very high tailwater levels only minor undulations occur (bottom row of Photo 4.1). Here, the non-modular flow pattern is obvious.

For flows at, or around, the modular limit, even an experienced hydraulic engineer will have difficulty in deciding whether flow is modular or not. The reason is that 'white water' at the surface may be caused either by a hydraulic jump or by a stable surface wave at subcritical flow (second and third rows of Photo 4.1). To aid in the visual detection of the modular limit, the following suggestions are made:

Modular flow, 25.47 m^3/s (18 May 1982)

At modular limit, 30.14 m^3/s (21 May 1982)

Photo 4.1. Flow over a broad-crested weir

Just beyond modular limit

Non-modular flow, 37.50 m^3/s (28 June 1982)

Photo 4.1. (cont.)

- Flumes with side contraction:

Whether flow through the flume is modular or non-modular can usually be seen from the way the jet leaves the flume throat. If the flow is modular, the flow pattern is almost symmetrical with respect to the centre line of the flume. If flow is non-modular, the jet moves either to the right or to the left, as shown in Photo 4.2. The side to which the jet moves can be influenced by disturbing or changing the flow pattern in the approach channel. By disturbing the flow pattern and observing the jet movements, one can test whether the flow is modular or not.

Photo 4.2. Non-modular flow; jet has moved to the left (Bos and Reinink 1981; photo University of Agriculture, Wageningen)

- Flumes (weirs) with bottom contraction:

With modular flow, the water surface at the downstream end of the throat curves downward. If the water surface curves upward at this location, flow is non-modular. At the modular limit, the water surface is horizontal (see Figure 4.9).

The modular limit is easier to detect if the end of the throat is marked on the side walls of the structure.

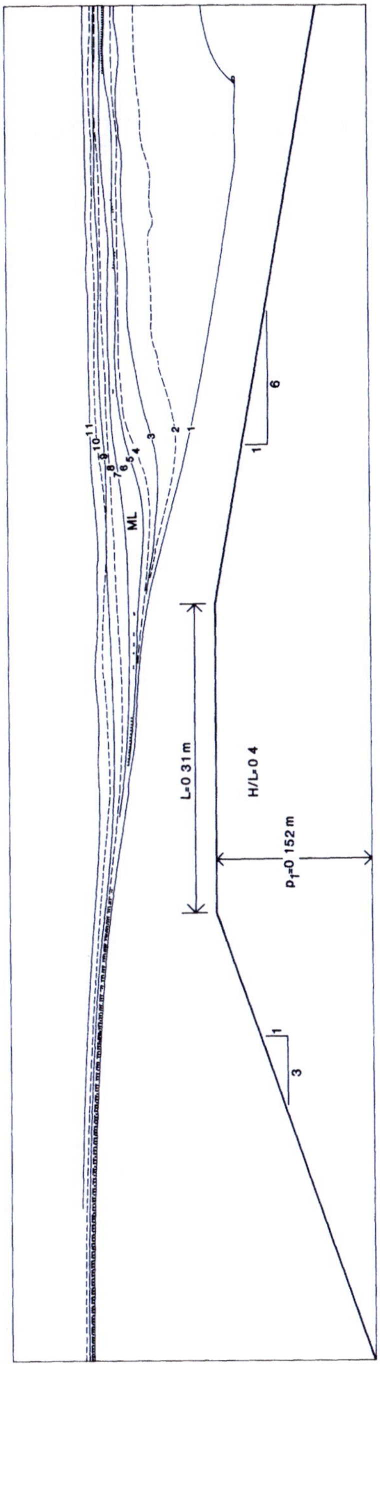

Figure 4.9. Visual detection of modular limit (two-dimensional flow)

5
Demands placed on a structure in an irrigation and drainage system

5.1 INTRODUCTION

If a structure is to function properly, the designer must list all the demands that will be made of the structure and match these demands to the properties of applicable structures. To aid in this matching process, the following demands will be discussed:

- Function of the structure (Section 5.2);
- Head loss for modular flow (Section 5.3);
- Range of discharges to be measured (Section 5.4);
- Error in the measurement of flow (Section 5.5);
- Restriction of backwater effect (Section 5.6);
- Capacity to transport sediment (Section 5.7).

A step-by-step design procedure is given in Section 5.8.

5.2 FUNCTION OF THE STRUCTURE

Irrigation and drainage demand three basic functions of a structure:

- Measurement of flow;
- Controlled regulation of flow;
- Control of upstream water level.

5.2.1 Measurement of flow rate

All broad-crested weirs and long-throated flumes are structures for measuring the rate of flow. To decide on what type of structure ought to be used, one must know for how long a period measurements are needed and how often they are to be taken. Together with information on the size and type of the channel in which the flow is to be measured, this will lead to the choice of:

- A portable and re-usable flume for the measurement of flow up to about 90 l/s (Clemmens, Bos, and Repogle 1984; see Photo 5.1).
- A temporary weir built of sufficiently durable sheet-material to last the anticipated measuring campaign (See Photo 5.2 for an example weir with a capacity of 1.5 m^3/s);
- A permanent weir or flume that measures the discharge in drainage canals and streams, or the flow rate in irrigation canals (see Photo 5.3).

As can be seen from Photos 5.1 to 5.3, the mere measurement of flow does not require a structure with movable parts. The upstream head is

Photo 5.1. Small portable RBC-flume in a drainage channel

Photo 5.2. Temporary weir, custom-built of sheet steel

Photo 5.3. A permanent broad-crested weir

measured with reference to the crest of the fixed sill. A processor can be installed with the head recorder to convert heads into flow rates. This processor can also integrate the flow rates to flow volumes per interval of time.

5.2.2 Controlled regulation of flow rate

When water is tapped from a river or reservoir or when an irrigation canal splits into two or more branches, a structure is needed that can both measure and regulate the flow. This calls for a weir whose sill is movable in vertical direction (Photo 5.4). With a near constant upstream water level, the upstream sill-referenced head can be adjusted so that the required flow passes over the weir.

If automated, the weir sill will be able to follow variations in the upstream water level and thus maintain a pre-set h_1 value and a corresponding flow over the weir.

Photo 5.4. Weir with movable sill at a canal bifurcation

5.2.3 Upstream water level control

The upstream water level can be controlled by a weir whose sill is either stationary or movable.

- Stationary sill: During the design phase, fixed values for y_{1min} and y_{1max} can be selected for the anticipated flow range. The control section is then designed so that these water depths occur at the corresponding Q_{min} and Q_{max}.
- Movable sill: Independent of the flow in the channel, the upstream water surface can be maintained at a certain level, or this level may be varied to suit a water management program.

As the present study is concerned only with flumes and with weirs that have stationary sills, weirs with movable sills will not be further considered.

5.3 HEAD LOSS FOR MODULAR FLOW

The available head loss at the measuring site follows from flow conditions in the downstream channel and the highest water level that is tolerated in the approach channel. This available head loss, at a given rate of flow, is a determining factor for the choice of the cross-sectional shape of the structure's control section and the expansion ratio of its diverging transition. On the other hand, the shape of the control section and of the downstream diverging transition influence the head losses at the required modular flow rates. These required head losses must be less than the head loss that is available at the measuring site, a requirement that must be met for the whole anticipated flow range.
This interrelationship between the available head loss, the required head loss, and the shape of the structure will be illustrated in the following example.

Example

- Given: A concrete-lined irrigation canal has a maximum design flow of 0.90 m^3/s. The canal has 1-to-1 side slopes, 0.60 m bottom width, a depth of 1.00 m, and a bottom slope of $s_b = 0.0009$.

- Required: Select a weir or flume to measure the flow range $Q_{min} = 0.10$ m^3/s to $Q_{max} = 0.90$ m^3/s with a maximum allowable upstream water depth $y_1 = 0.90$ m.

Use the Manning equation to determine the relationship between Q and y_2:

$$Q = A_2 \frac{1}{n} R_2^{2/3} s_b^{1/2} \quad (5.1)$$

where A_2 and R_2 are functions of the water depth, y_2, and 1/n is expressed in $m^{1/3}/s$.
The hydraulic resistance, n, determines the water depth, y_2, at which a certain flow passes through a channel with a certain cross-section and bed slope. The value of n is not expected to be constant because it is influenced by a variety of factors: erosion, sedimentation, deposition of trash, growth of weeds or algae, deterioration of lining, lack of maintenance, and so on. To illustrate the influence of some of these factors on the (tail)water depth, let us assume n = 0.014, a value commonly used for a concrete-lined irrigation canal under field conditions (USBR 1957; Chow 1959). With this n value, the flow rate can be calculated for each water depth. The resulting y_2-Q curve is shown in Figure 5.1 as the dashed line.

The concrete bottom of the canal, however, may be covered with sediment. If so, the n value will be higher. If a 0.05 m sediment layer covers the bottom and n = 0.016, the y_2-Q curve becomes the dotted line of Figure 5.1. If the hydraulic roughness is significantly higher because of neglected maintenance, the n value can become 0.020, resulting in the solid line of Figure 5.1. Such a line could be considered a conservative estimate of water levels for the given range

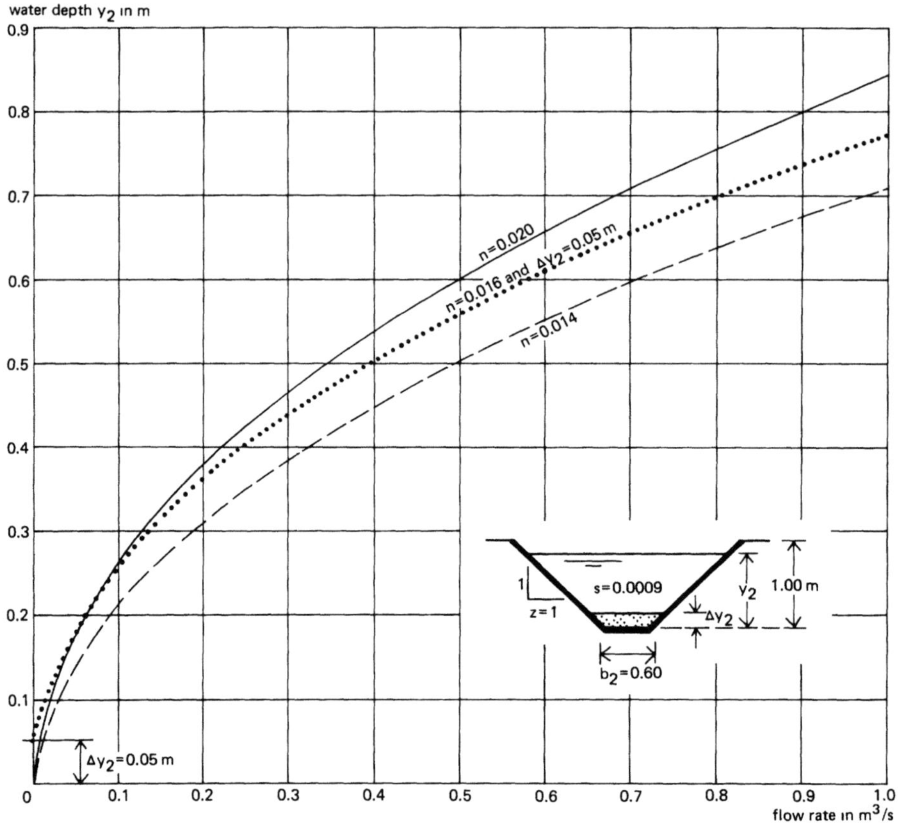

Figure 5.1 Range of possible water levels for a given flow in an example canal

of flow rates. To study whether flow over a structure will be modular, it is recommended that such a conservative upper limit be used for the possible tailwater levels. In this context, it should be noted that tailwater levels can also be influenced by the operation of gates, by backwater effects, and so on.

In practice, it is sufficient to make a conservative estimate of the tailwater levels at the minimum and maximum flows that are to be measured and to check the structure for modular flow at these extreme flow rates. For this example, the solid line of Figure 5.1 is used. The maximum flow rate, $Q = 0.90$ m^3/s, corresponds with $y_2 = 0.80$ m.

The following step in the procedure is to find, by trial and error, the shape of the control section, the sill height, and the expansion ratio of the diverging transition, so that, at any flow rate to be measured:

$$p_2 + (ML \times h_1) > y_2 \qquad (5.2)$$

As was shown in Chapter 4, the modular limit, ML, depends, amongst other things, on the downstream flow velocity and on the expansion ratio (EM to 1) of the diverging transition.

To meet the condition of Eq. 5.2, a wide control section with a high downstream sill (high p_2 value) and low head (h_1) is appropriate. An example of such a structure is shown in Figure 5.2A. The head-discharge relationship for this weir, with $p_1 = p_2 = 0.45$ m and $b_c = 1.50$ m, must then be determined as was explained in Chapter 3. This reveals that $Q_{max} = 0.90\ m^3/s$ at a head $h_1 = 0.42$ m. Figure 5.2A shows the h_1-Q relationship of the weir and that between the upstream water depth, $y_1 = h_1 + p_1$, and Q. For Q = 0.90 m^3/s, it shows that $y_1 < 0.90$ m, thus meeting the postulated demand. With the method explained in Chapter 4, the corresponding modular limit, ML = 0.89, is found. The maximum allowable tailwater depth for modular flow is then $y_{2max} = 0.45 + (0.89 \times 0.42) \approx 0.82$ m, which is 0.02 m above the depth $y_2 = 0.80$ m.
For other rates of flow, the related values of head (h_1) and modular limit (ML) can be calculated, again yielding an allowable value y_{2max}. Figure 5.2A also shows a curve of y_{2max} versus Q.

For a continuous channel bottom, $p_1 = p_2$, as in this example, the difference between the y_1 curve and y_{2max} curve yields the required head loss. Between the y_1 curve and y_2 curve, the actual head loss is read.
For modular flow to occur, the actual head loss must be greater than the required head loss. Thus, the y_{2max}-Q curve should be above the y_2-Q curve of Figure 5.2A.

As shown above, the margin between the required and actual head loss is 0.02 m at $Q_{max} = 0.90\ m^3/s$, but it increases with decreasing flow rate.

In this example the sill height $p_1 = p_2 = 0.45$ m and the actual head loss equals 0.29 m for $Q_{min} = 0.10$ m^3/s, whereas only 0.04 m is required. If the value of h_1 at Q_{min} is too low to meet the accuracy requirement with which Q_{min} is to be measured, the bottom of the control section must be narrowed and lowered so that h_{1min} will increase as will be shown in Section 5.5.
To obtain an optimal accuracy at all flow rates, a flume can be designed with the control section shaped in such a way that the y_1-Q curve follows the y_2-Q curve throughout the anticipated flow range. Such a flume, which has a complex trapezoidal cross-section, is shown in Figure 5.2B ($p_1 = p_2 = 0$). As can be seen, y_1 does not exceed 0.90 m if Q equals 0.90 m^3/s.

In this flume, with $p_1 = p_2 = 0$ and $b_c = 0.10$ m, the condition of Eq. 5.2 can only be met if the modular limit, ML, is high.
Figure 5.2B shows two diverging transitions:

1: EM = 0. An abrupt diverging transition. The modular limit at Q = 0.90 m^3/s equals ML = 0.86 and the maximum tailwater depth is $y_{2max} = 0.76$ m. This value is below the actual tailwater depth so that the flow is non-modular. In fact, Figure 5.2B shows that flow becomes non-modular above 0.15 m^3/s.

2: EM = 6. If a diverging transition with EM = 6 is added to the same flume throat, the modular limit increases to 0.91 and thus for Q = 0.90 m^3/s $y_{max} = 0.82$ m, which is above the actual y_2 value. For lower rates of flow, too, the y_{2max} curve lies above the y_2-Q curve and flow remains modular at all anticipated rates.

A comparison between the weir ($p_1 = 0.45$ m) of Figure 5.2A and the flume (p = 0, EM = 6) of Figure 5.2B shows the flume to be a relatively costly construction. A significant part of its cost is related to the criterion of accuracy, which implies that the sill-referenced head must be as high as possible at low flows. This criterion requires that $p_1 = 0$ and $b_c = 0.10$ m. If this criterion can be relaxed, the lower trapezoid of the complex control section can be replaced by its equivalent singular rectangular control section (Figure 5.2C). A comparison of Figures 5.2B and 5.2C shows that the two y_1-Q curves almost coincide whereas the design of the Figure 5.2C structure is less complicated and less costly.

The flume of Figure 5.2C has a diverging transition with EM = 6 flare, which is needed for modular flow to occur at this low p_1 value.
A structure that has less than 0.10 m backwater effect at the maximum flow rate (y_1 < 0.90 m at Q = 0.90 m^3/s and y_2= 0.80 m), and does not need a diverging transition, can be realized only if p_2 increases at the cost of h_1 (see Eq. 5.2). If the head, h_1, decreases, however, the width of the control section must increase to obtain sufficient discharge capacity at y_1 = 0.90 m.
The lowest possible sill that can function with an abrupt diverging transition (EM = 0) is shown in Figure 5.2D. With respect to the design of Figure 5.2C, the sill has been raised from 0.15 to 0.30 m and the width of the control section has been increased from 0.80 to 1.10 m. Because of the increase of p_1 by 0.15 m at the cost of h_1, the measurement of flow will be less accurate (see Section 5.5)

Another dimension that varies with the upstream sill-referenced head in the four designs of Figure 5.2 is the throat length, L. In this example the ratio H_{1max}/L < 0.75, so that L varies from 0.60 m in Design A to 1.20 m in Design B. This length is another factor that influences the construction cost of the weir or flume.

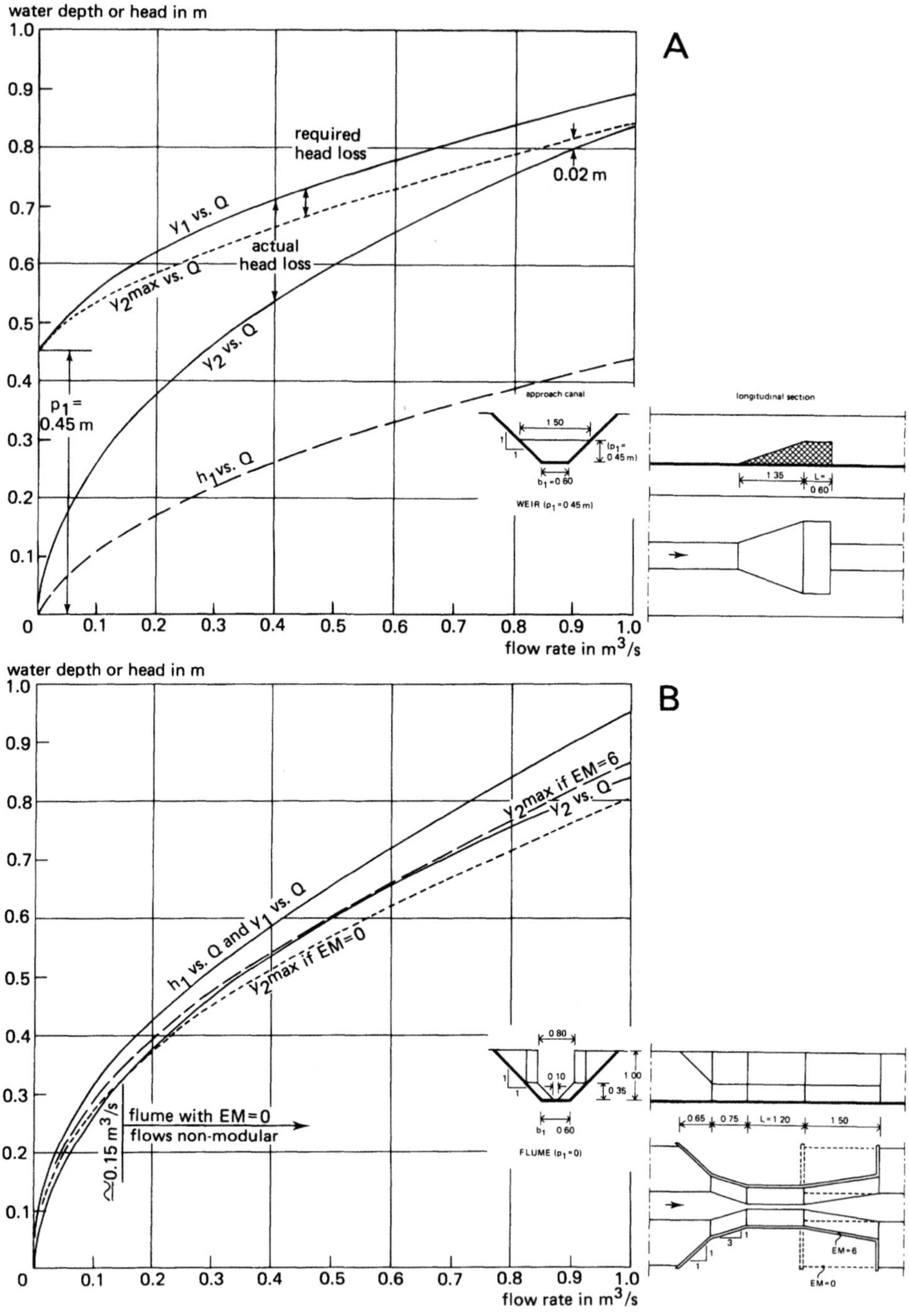

Figure 5.2. Alternative weirs and flumes for flow measurement in a concrete-lined canal

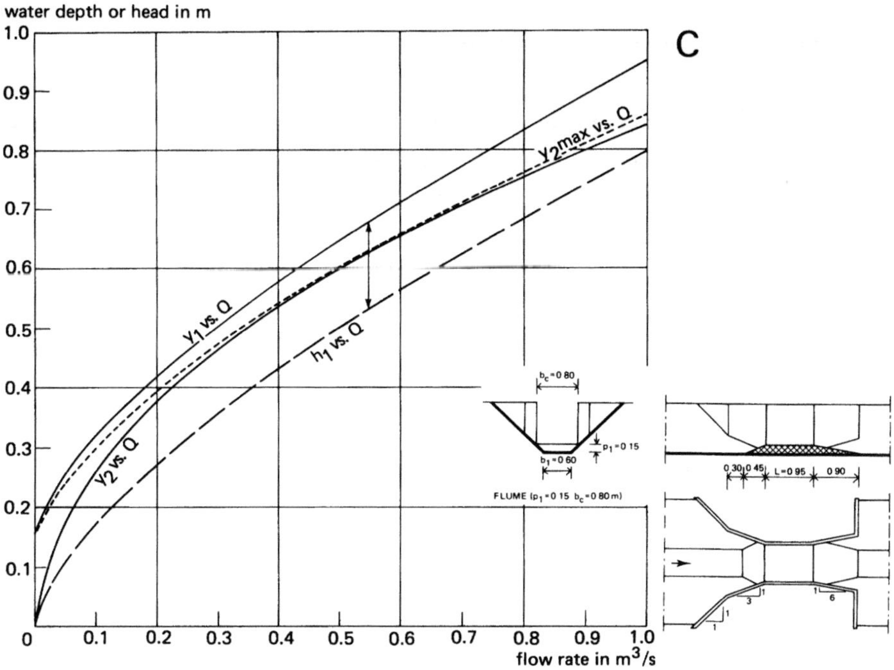

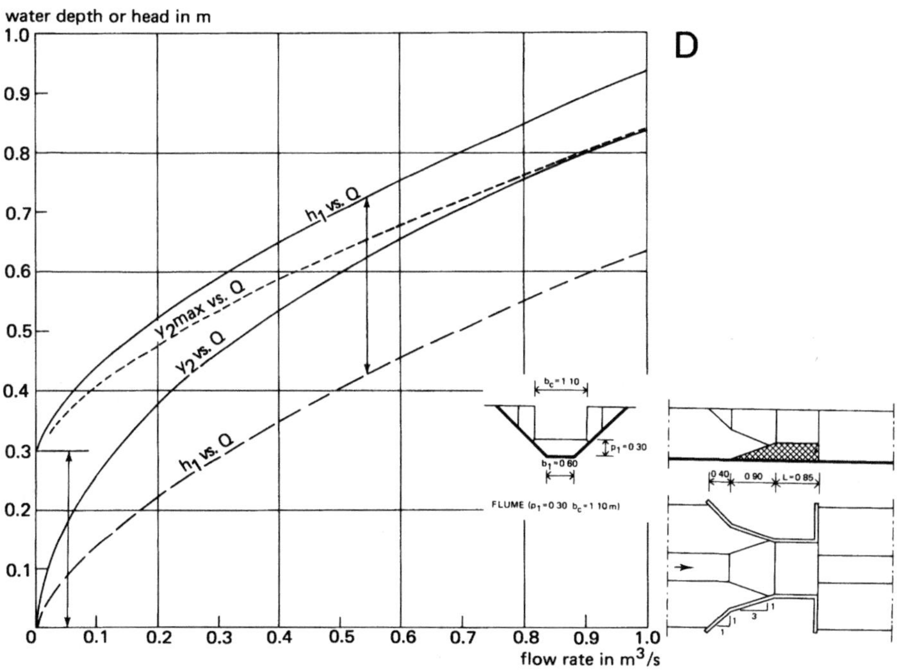

Figure 5.2. (cont.)

5.4 RANGE OF DISCHARGE TO BE MEASURED

Generally, the flow rate in an open channel varies with time. The range between Q_{min} and Q_{max} through which the flow is to be measured, strongly depends on the type of channel in which the structure is placed. Irrigation canals, for example, convey a considerably narrower range of discharges than do natural streams. The anticipated range of discharges to be measured can be classified by the ratio:

$$\gamma = Q_{max}/Q_{min} \qquad (5.3)$$

For rectangular, parabolic, and triangular control sections, the head-discharge relationship can be expressed in the form:

$$Q = C_d K H_1^{\,u} \qquad \text{(Eq. 4.3)}$$

The discharge coefficient, C_d, can be derived from Figure 3.16 for $0.1 < H_1/L < 1.0$. For Q_{max} the ratio $H_1/L = 1.0$ and $C_d = 1.025$; for Q_{min} the ratio $H_1/L = 0.1$ and $C_d = 0.930$. It follows that the corresponding range of flows that can be measured by a certain control section is:

$$\gamma = \frac{Q_{max}}{Q_{min}} = \frac{1.025}{0.930}\, 10^u = 1.1 \times 10^u \qquad (5.4)$$

As was discussed in Chapter 4, the exponent, u, depends on the shape of the control section perpendicular to the direction of flow. For a rectangular control section, $u = 1.5$, and thus $\gamma = 35$; for a triangular control section, $u = 2.5$ and $\gamma = 350$.
For other shapes of the control section (trapezoidal, complex, etc.) the exponent u varies from 1.6 to 2.4 depending on the ratio B_c/b_c. The actual u value can be determined from Eq. 4.4 or 4.5.

In irrigation canals, the ratio $\gamma = Q_{max}/Q_{min}$ is well below 35 and a rectangular (or trapezoidal) control can measure all anticipated flow rates. In drainage canals and natural streams, $\gamma = Q_{max}/Q_{min}$ usually

exceeds 35, so that narrow-bottomed trapezoïdal, triangular, or complex control sections are appropriate.
If $\gamma = Q_{max}/Q_{min}$ exceeds 350, a specially shaped control section, with a notch in its bottom, can be used. Although a rating table can be produced for the upper and lower ranges of head in such a control shape (see Section 3.4), such structures are commonly calibrated in a hydraulic laboratory.

5.5 ERROR IN THE MEASUREMENT OF FLOW

The allowable error in the measurement of flow has an important bearing on the relation between the (average) width and depth of the control section of a flume or weir.
The accuracy to which a flow rate can be measured depends on:

- The accuracy with which a rating table can be made for the structure. The percentage error in the discharge coefficient is given in Eq. 3.62 as $X_c = \pm (3 \left|H_1/L - 0.55\right|^{1.5} + 4)$ per cent;
- The accuracy with which the upstream sill-referenced head, h_1, can be determined. The resulting percentage error in h_1 is denoted as X_{h1};
- The exponent, u, in the head-discharge equation (see Section 4.2.1.).

If the sill-referenced elevation of the head measurement device has been determined with sufficient accuracy, so that this source of systematic error is removed, the value of X_Q for modular flow is:

$$X_Q = \left[X_c^{\,2} + (uX_{h1})^2\right]^{0.50} \tag{5.5}$$

This X_Q should not exceed a critical value to be chosen by the user of the flow data. Usually one value is chosen for X_{Qmin} and another for X_{Qmax}. The relative magnitude of both values depends on the shape of the hydrograph, i.e. on the rates of flow that pass the structure during high or low flow periods. For a given value of X_Q and derived values of X_c (Eq. 3.6.2) and u (Eq. 4.4 or 4.5), the allowable error in h_1 can be calculated from:

$$X_{h1} = \left[(X_Q{}^2 - X_c{}^2)/u^2\right]^{0.50} \qquad (5.6)$$

For the h_1 value, which is related to the considered flow rate, the maximum allowable reading error, Δh_1, in h_1 can be calculated from:

$$\Delta h_1 = (h_1 \times X_{h1})/100 \qquad (5.7)$$

Usually, the chosen value of X_{Qmin} will lead to a smaller Δh_1 value than the one that results from the postulated X_{Qmax} value. As a result, the Δh_1 value at Q_{min} becomes a determining factor in the design of the cross-section of the control and in the selection of a head-measuring device.

In practice, the total error in the sill-referenced head consists of two components: one due to a faulty reference level and the other, Δh_1, due to reading or registration errors. The systematic error due to a faulty reference level depends on the stability of the head-measuring device with respect to the sill reference point, on the care with which the procedure of zero-setting the device to the reference point is done and on the growth of algae and the deposit of sediment in the control. If the structure is properly maintained, the related error in h_1 can be limited to 1 mm (Brakensiek, Osborn, and Rawls 1979; Bos, Replogle, and Clemmens 1984). Any attempt at accuracy will fail if the structure is not scrupulously maintained.
The reading error, Δh_1, which is related to the head-measuring device, can be based upon information from instrument manufacturers, laboratory research (Landbouwhogeschool, Wageningen 1968), and field experience (Bos, Replogle, and Clemmens 1984).

Once the head-measurement method has been decided upon, Δh_1 can be determined. The requirement for h_{1min} for keeping the error below Q_{min} is then:

$$h_{1min} = 100\ \Delta h_1/X_{h1} \qquad (5.8)$$

At this low head, the bottom width of the control section can be approximated with a simplified modification of Eq. 3.18 (b_c and h_{1min} in metres, Q_{min} in cubic metres per second):

$$b_c \simeq Q_{min}/1.7\, h_{1min}^{1.5} \tag{5.9}$$

Equation 5.7 shows that for a certain value of the reading error, Δh_1, the percentage error in the sill-referenced head will decrease as the head, h_1, increases; it follows from Eq. 5.5 that the percentage error in the matching flow rate decreases accordingly.

5.6 RESTRICTION OF BACKWATER EFFECT

The water depth, y_1, in the channel upstream of a weir or flume equals the 'undisturbed' water depth, plus a backwater effect created by the structure. To avoid overtopping of the channel at Q_{max}, and to minimize sedimentation problems, this backwater effect is limited, y_1 not being allowed to exceed a pre-determined value at maximum flow. The Q_{max} at this water depth, y_{1max}, can be characterized by the Froude number at the gauging station:

$$Fr_1 = \frac{v_1}{\sqrt{gA_1/B_1}} \tag{5.10}$$

To provide an upstream water surface that is sufficiently stable to allow the sill-referenced head to be measured, Fr_1 should not exceed 0.5.

In the control section, flow is critical:

$$Fr_c = \frac{v_c}{\sqrt{gA_c/B_c}} \tag{5.11}$$

To create $Fr_c = 1.0$, the flow must be accelerated towards the control section; this is done by contracting the channel. The degree of contraction can be described by the ratio A^*/A_1, where A^* is the area of the control section below the upstream water level (see also Section

3.6). As shown in Figure 5.3, the upper limit of the area ratio, A^*/A_1, increases with increasing Fr_1 value.

The empirical curve of Figure 5.3 is the envelope of the calculated relationship between Fr_1 and A^*/A_1 for modular flow in a variety of structures. To illustrate the character of this relationship, data points are shown for four control sections: the rectangular control section of Figure 5.2D, a triangular control ($p_1 = 0$, $\Theta = 90°$) in the same approach channel, the ($p_1 = 0.25$ m) trapezoidal control of Figure 5.5, and the rectangular-throated weir of Figure 5.6.

To determine the required area A*, one must know the dimensions of the approach channel and the allowable water depth, y_{1max}, at Q_{max}. Upon calculation of the related value of Fr_1, the corresponding ratio A^*/A_1 can be read from Figure 5.3. If a structure is designed with a lower A^*/A_1 ratio, this structure may cause more than the postulated backwater effect.

If a structure is designed with a higher A^*/A_1 ratio than that derived from Figure 5.3, the upstream water depth at Q_{max} will be lower than y_{1max}.

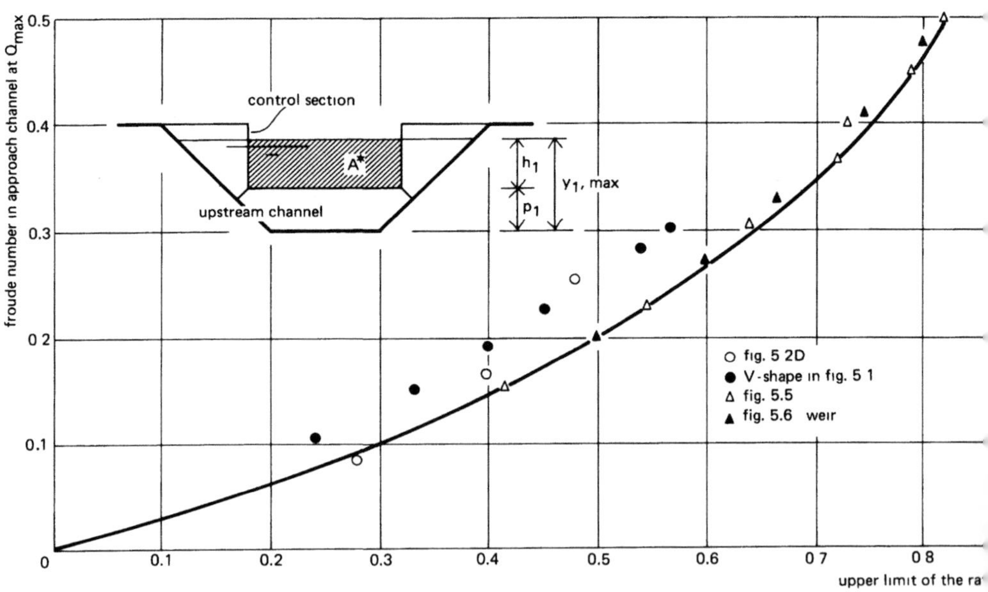

Figure 5.3. Approximate upper limit of A^*/A_1 as a function of Fr_1

5.7 SEDIMENT TRANSPORT CAPACITY

5.7.1 Design rule

The most appropriate method of avoiding sediment deposition in the channel reach upstream of the structure is to avoid a decrease in the hydraulic gradient. To achieve this, the structure should be designed in such a way that it does not create a backwater effect. With respect to the average approach channel bottom, this means that the Q versus ($h_1 + p_1$) curve of the control must coincide with the water depth-discharge curve (Q versus y_1 curve) of the channel upstream of the structure. This near coincidence should occur for those anticipated flows that are liable to transport bed-load.

The effect of this design rule on the shape of the weir or flume can be illustrated by considering a concrete-lined irrigation canal which transports sediment that may originate from the diverted river, from runoff into the canal, and from wind erosion. The dimensions of the canal are $b_1 = 3.0$ m, $z_1 = 1.5$, $d_1 = 1.20$ m, and s = 0.00125. The design flow rate of the canal is $Q_{max} = 6.1\ m^3/s$, which is transported at a water depth of about $y_1 = 1.0$ m, so that the related maximum value of the Froude number is 0.50. For such irrigation canals (see also Photo 4.1) the rougness of the concrete side slopes is less than that of the covered bottom. As a result, the Manning n value increases with decreasing water depth.
The water depth versus flow-rate relationship of this canal is shown in Figure 5.4. It can be approximated by:

$$Q \simeq 6.1\ y_1^{2.0} \qquad (5.15)$$

with y_1 in metres and Q in cubic metres per second.

The above design rule requires a drop in the canal bottom that is sufficient to guarantee modular flow (here taken at 0.20 m), and a control section that has a capacity of 6.10 m^3/s at the undisturbed water depth, $y_1 = p_1 + h_1 = 1.00$ m. The simplest control that meets this demand is a horizontal weir sill of $p_1 = 0.25$ m height

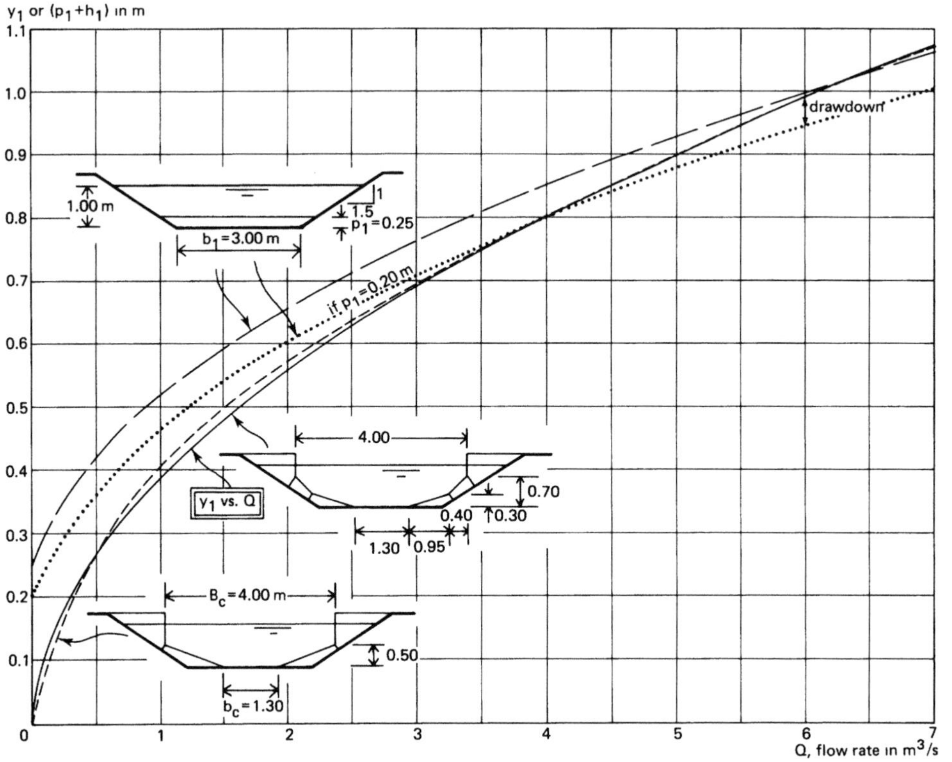

Figure 5.4. y_1 versus Q curve for a concrete lined irrigation canal and $(p_1 + h_1)$ versus Q curves of suitable structures

(see Figure 5.4). At 50% of the design discharge, this weir causes a backwater effect of 0.07 m. This effect can further be reduced if a weir sill height of p_1 = 0.20 m is selected. The y -Q curve of this weir is shown as a dotted line in Figure 5.4. It reveals a backwater effect of only 0.02 m at 3.0 m^3/s, and a drawdown of 0.04 m at 6.1 m^3/s. With flows below 3 m^3/s, the backwater effect increases, which may cause sedimentation if the flow velocity, v_1, in the undisturbed channel is sufficient to transport sediment. For lower flows, however, the water depth in the undisturbed channel, and the related v_1, will decrease, and eventually the transport of bed load will cease.

If no backwater effect is allowed for flows below 6.1 m^3/s, a flume with zero bottom sill ($p_1 = 0$) is the only solution. Because of the power ≈ 2 of y_1 in Eq. 5.15, the shape of the control section must approach that of a parabola (u = 2.0) with a focal distance of $f_c = 1.00$ m (see Figure 3.12). The h_1–Q curve of the complex trapezoidal control, with double break in the sides, coincides with the y_1–Q curve of the channel (see Figure 5.4). To economize on the construction cost of the flume, the control section can be simplified to a trapezoid with a single break in the sides. As shown in Figure 5.4, the backwater effect of such a flume is negligible. Figure 5.5. shows that the weir is simpler, and therefore cheaper, than the flume.

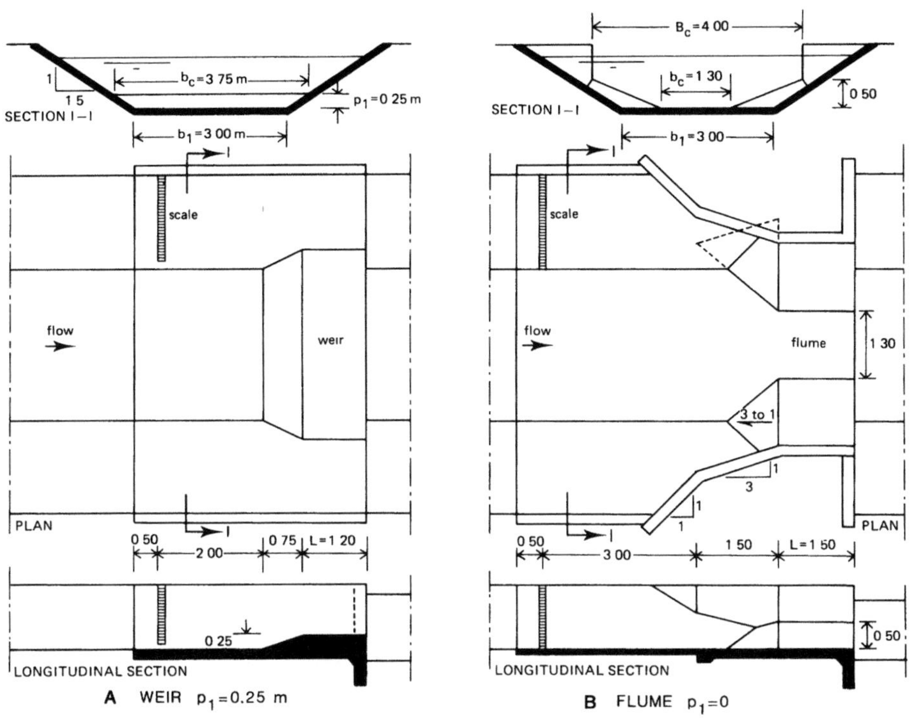

Figure 5.5. Dimensions of a weir and flume

5.7.2 Laboratory experiments

At first glance, the flume of Figure 5.5 appears superior to the weir in its capacity to pass sediments. Available field experience, however, reveals no sediment problems with weirs that meet the design rule on the restriction of the backwater effect of Section 5.7.2; all sediment that could be transported by a channel passed the weirs.

To evaluate this field experience and to generate systematic information on the influence of the transport of sediment on the h_1-Q relationship, some exploratory laboratory tests were run to compare a broad-crested weir with a long-throated flume. The dimensions of the two structures are shown in Figure 5.6.

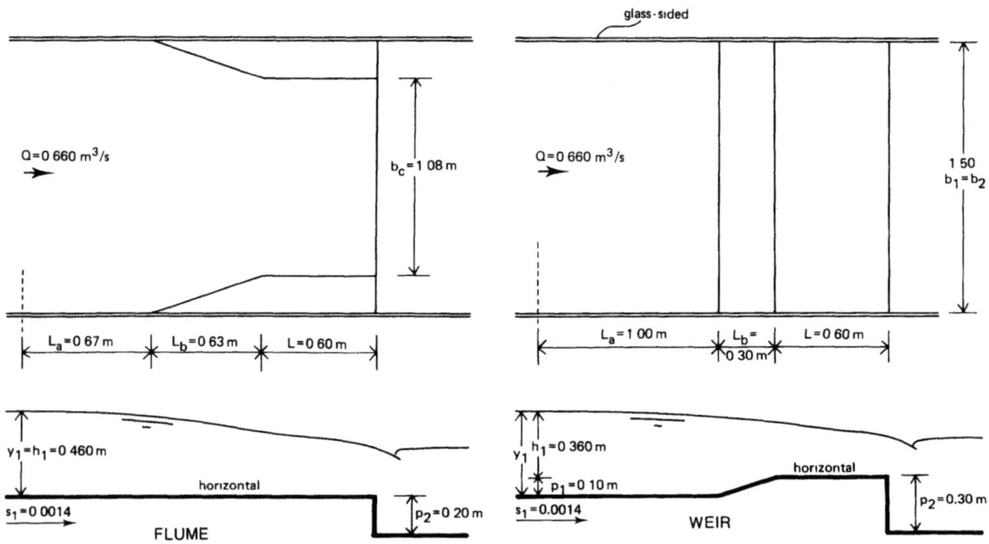

Figure 5.6 Dimensions of the tested weir and flume

Both structures were placed in a glass-sided laboratory channel (b_1 = b_2 = 1.50 m, z = 0, and s_1 = 0.0014). They were dimensioned in such a way that, without sediment, the Froude number in the approach channel was equal to 0.45, so that Q = 0.660 m^3/s and $y_1 = h_1 + p_1$ = 0.460 m. As a result, flow conditions upstream of the gauging stations of the weir and flume were identical. Figure 5.7. shows that they remained identical with increasing sediment load. The

tailwater level was sufficiently low for flow to be modular.
In these exploratory tests, the sediment diameter and Froude number were so chosen that, after a change in the regime, steady-state conditions were reached within an acceptable period.
While flow remained constant at Q = 0.660 m^3/s, the sediment load was increased for each test run (D_{10} = 0.00071 m, D_{90} = 0.00084 m). During each run, the sediment transport was allowed to stabilize before final data were measured on approach channel elevation, sill-referenced head, h_1, and total energy head, H_1.

During the tests, the total sill-referenced energy head remained stable at H_1 = 0.400 m for the weir and H_1 = 0.506 m for the flume. As a result of the increasing sediment transport, the average elevation of the approach channel bottom increased as shown in Figure 5.7. No deposit of sediment was observed on the weir crest or in the flume throat, so that the reference level did not change (Photo 5.5).

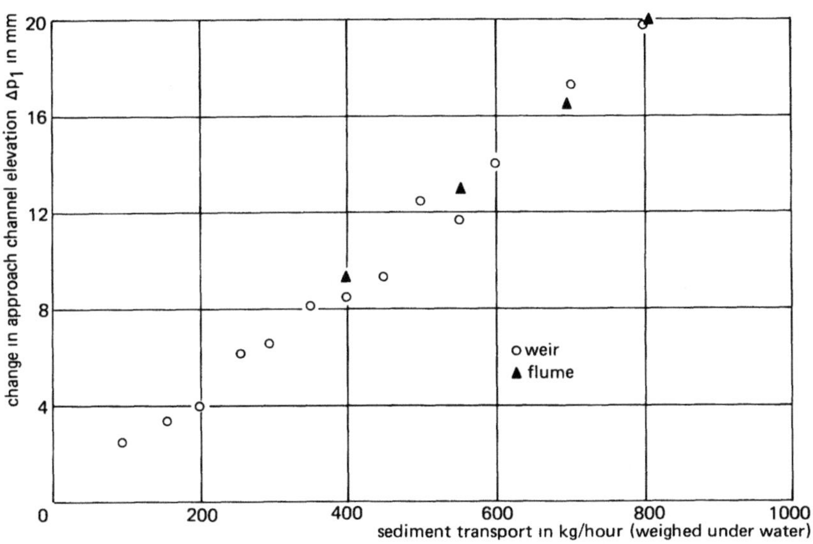

Figure 5.7. Change in approach channel bottom elevation with sediment load (Bos and Wijbenga, in prep.)

Photo 5.5. Weir flow with Q = 0.660 m^3/s and a sediment load of 800 kg/hour (photo Delft Hydraulics Laboratory)

The resulting decrease of the water depth at the gauging stations of the weir and flume caused $v_1^2/2g$ to increase at the cost of h_1 (see Section 3.2). The measured decreases of h_1 are shown in Figure 5.8 as a function of the sediment load. The variation of h_1 data is due to the changing elevation of the sand dunes. The related 95% confidence limits are shown.

The decrease of sill-referenced head because of an increase in the approach velocity can be forecast accurately by applying the theory of Section 3.6. Figure 5.8 shows that, for the same sediment transport, the decrease in h_1 is significantly greater for the weir than for the flume. Because the h_1 value of the weir is 0.10 m less than that of the flume, the percentage reduction of h_1 is even more significant, especially because of the exponent u = 1.5 of h_1 in the h_1-Q equation

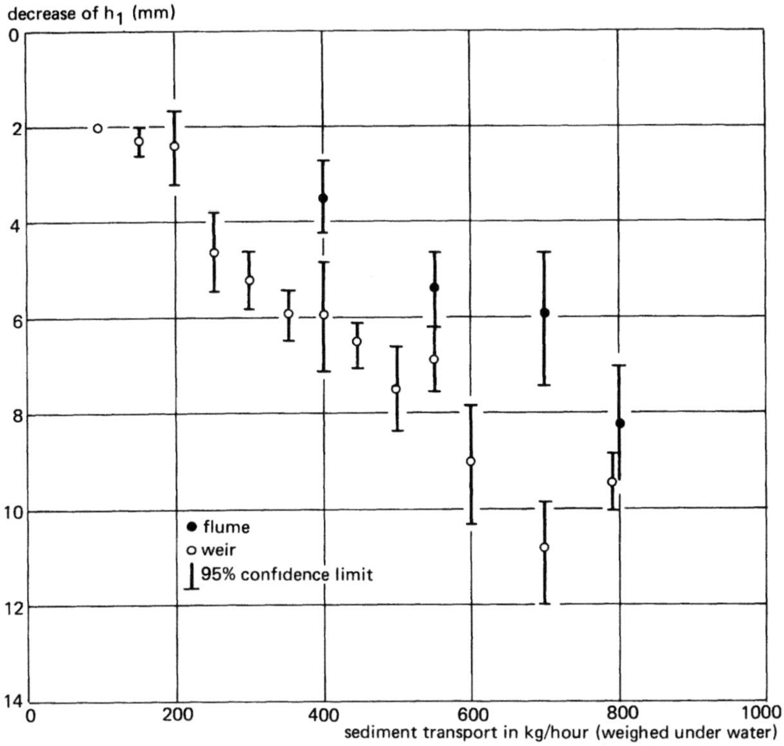

Figure 5.8. Decrease of sill-referenced head because of change in p_1 related to sediment transport (Bos and Wijbenga, in prep.)

of both structures. For example, a sediment transport of 700 kg/h causes an error of 4.9% in the measured weir flow against an error of 2.0% in that of the flume. Hence, in sediment-laden flow, a flat-bottomed flume performs better than a weir. An example of such a flume, with minimum backwater effect (see Section 5.7.2), is shown in Photo 5.6. This prototype flume has performed satisfactorily since 1970.

Photo 5.6. Flat-bottomed flume in the High School Wash, Tucson, Arizona

The results of the laboratory experiments can be applied to a prototype by use of (de Vries 1973):

$$n_{v1}^{2} = n_{Da} \tag{5.12}$$

where:

n_{v1} = velocity scale;
n_{Da} = particle diameter scale.

This scale condition on the sediment-transport process usually influences the Froude scale condition:

$$n_{v1}^{2} = n_{Fr1}^{2} n_{y1} \tag{5.13}$$

where:

n_{Fr1} = Froude number scale;
n_{y1} = water depth scale.

Because laboratory models are usually designed with $n_{y1} > n_{Da}$, the scale conditions of Eqs. 5.12 and 5.13 can only be met if $n_{Fr1} < 1$. Hence, the Froude number in the model is larger than that in the prototype.

A prototype that would match the laboratory tests is a channel with a water depth of 2 m and an average bed-load diameter of 3.5 mm. Such a channel could be a river or a large irrigation canal. The test data could also apply to a smaller concrete-lined irrigation canal, which transports sediments that originate from wind inblow.
The results of these exploratory tests warrant further research, being applicable to channels of shallower depth. In such experiments, however, considerably more time will be required before steady-state conditions are attained. Research on structures in trapezoidal channels is also recommended.

5.8 DESIGN PROCEDURE FOR A STRUCTURE

In the above sections, the designer of a long-throated flume or broad-crested weir has been given a number of guidelines to select the dimensions for a structure in a certain channel. The design procedure is an iterative process which leads from a detailed description of the measurement site to a number of suitable structures. The number of structures that can serve a given site satisfactorily depends on the channel in which the structure is to be placed and on the postulated hydraulic demands.

The steps in the design procedure are:

1. Obtain data on the hydraulic dimensions and boundary conditions of the channel at the measuring site and make a conservative estimate of the Q-y_2 relationship.

2. Depending on the period that a structure is to be used and on Q_{max}, decide whether a portable, temporary, or permanent

structure is needed, and, if a weir, whether it should have a stationary or movable sill (see Section 5.2).

3. Decide on the allowable error in the measurement of Q_{min} and Q_{max}.

4. Determine the ratio, $\gamma = Q_{max}/Q_{min}$, and tentatively select the shape of the control section (rectangular, trapezoidal, complex, etc.). The actual dimensions will be selected in Step 6 (see Section 5.4.)

5. Determine the maximum permitted water depth in the upstream channel (y_{1max}) and calculate the related values of A_1 and Fr_1. Use Figure 5.3 to find the ratio A^*/A_1. Calculate A^*.

6. Make a first tentative choice of the dimensions of the control section, using the shape of Step 4 and the area A* of Step 5.

7. Make a preliminary estimate of H_{1min} and H_{1max}. Determine the flow-wise length of the throat, L, in such a way that $0.1 < H_1/L < 0.1$. If the available head loss is a limiting factor, try $H_1/L < 0.5$.

8. Use Chapter 3 to calculate h_{1min} and h_{1max} for the given range of flows to be measured by the control section of Step 6.

9. Use Section 5.5 to calculate the allowable reading error, Δh_{1min} and Δh_{1max}, for the given Q_{min} and Q_{max}. If no practical head measuring device is available to meet the demanded value of Δh_1, reduce the width of the control section (see also Eq. 5.9) and return to Step 7. Note that the area A* of this new control section must again equal the value of Step 5.

10. Use the condition $p_2 + (ML \times h_1) > y_2$ to determine the values of h_{1min} and h_{1max} that are required for modular flow at Q_{min} and Q_{max} (see Eq. 5.2). For a first trial, use preliminary values of

ML = 0.85 for a structure in an ongoing channel; for a structure dispensing into a wide channel or reservoir, use ML = 0.60. In a final phase of the design process, the modular limit (ML) must be estimated with the procedure of Section 4.2.5.

11. Compare the value of h_{1min} of Step 8 with the h_{1min} value of Step 10.
If the head required to measure Q_{min} (Step 8) is equal to, or greater than, the head required for modular flow (Step 10), proceed to Step 12. Otherwise, raise the invert of the weir crest or flume throat.
To meet the demand on A* of Step 5, make the control section wider. With these new dimensions of the control section, return to Step 7.

12. Compare the value of h_{1max} of Step 8 with the h_{1max} of Step 10. If the head required for modular flow (Step 10) is less than the head required to measure Q_{max} (Step 8), a structure can be selected that will meet all of the design criteria and will operate accurately. This, however, does not guarantee the best or most efficient structure. If structure is accepted, proceed to Step 13.
If the head required for modular flow is greater than that required to measure Q_{max}, a structure cannot be selected that will perform as desired, regardless of the width of the control section. In this case, the designer has various options:
a) Increase the allowable measurement error;
b) Use a more accurate head detection method;
c) Increase the allowable upstream water level by raising the canal walls or reducing the freeboard requirements;
d) Reduce the required head loss by adding a diverging transition;
e) Choose a location where more head loss is available.
Then repeat Steps 1 through 11, as appropriate.

13. Return to Step 10 to ensure that the modular limit at Q_{min} and Q_{max} was estimated with the procedure of Section 4.2.5. If so, proceed to Step 14.

14. Return to Chapter 3 to calculate the h_1-Q relationship of the structure.
    Plot this relationship with the y_1-Q and y_2-Q curves in one graph (see Figure 5.2).

The above design procedure is an iterative process, with a number of trials before a final design is selected. Although this procedure appears to be fairly complex, the various iterations converge rapidly. The only difficult part of the procedure is estimating the flow conditions prior to the placement of the structure.

Summary

Vital for effective water management are structures that can measure the rate of flow in a wide variety of canals and streams. Chapter 1 of this thesis introduces the long-throated flume and the broad-crested weir; it explains why this family of structures can meet the boundary conditions and hydraulic demands of most measuring sites.
Chapter 2 records the history of long-throated flumes and broad-crested weirs. It describes how the hydraulic theory of flumes and weirs, and their design, developed separately over the last hundred years. The chapter concludes by reporting recent attempts to develop a generally valid theory for any long-throated flume or broad-crested weir in any channel.

The remainder of the thesis explains the steps that were taken to arrive at a procedure (Section 5.8) that can yield the hydraulic dimensions and the rating table of the appropriate weir or flume to measure the flow rate in any channel. The major steps taken cover the hydraulic theory of the flow of water through control sections of different shapes and dimensions (Chapter 3), the theory and procedure of estimating the head loss required for modular flow (Chapter 4), the boundary conditions of the channel in which flow is to be measured, and the demands placed on the structure with respect to the range and accuracy of its measurements (Chapter 5).

Chapter 3 gives two basic equations: $Q_i = A_c[2g(H_1 - y_c)]^{0.50}$ (Eq. 3.12) and $y_c = H_1 - A_c/2B_c$ (Eq. 3.13), which are valid for critical flow of an ideal frictionless liquid in any long-throated flume and broad-crested weir. With these equations, the H_1 versus Q_i relationship can be derived for those control sections which have well-defined expressions for their A_c and B_c. Section 3.3 presents this relationship for singular rectangular, triangular, trapezoidal, parabolic, circular, and truncated-circular control shapes, and for the composite truncated-triangular and U-shaped control sections. For other composite shapes, a procedure by which the H_1 versus Q_i relationship can be found is given in Section 3.4. This procedure is based on the property of these flumes and weirs that the ideal flow rate, $Q_i = g^{0.50} A_c^{1.50} / B_c^{0.50}$, is the same through all shapes of control sections that flow with the same B_c and have the same area A_c. It is explained how the H_1 versus Q_i relationship of an equivalent singular control shape can be used for certain composite control sections.

To convert the ideal into the real flow rate, a discharge coefficient is used: $C_d = Q/Q_i$ (Eq. 3.59). This C_d value is assumed to be influenced mainly by friction along, and streamline curvature above, the flume throat or weir crest (with length L). An analysis of laboratory data on 105 flumes and weirs of different shapes and sizes and made from different construction materials produced a curve that gives the average C_d value as a function of H_1/L. Within the limits of application, $0.1 \leqslant H_1/L \leqslant 1.0$, the C_d value of any long-throated flume or broad-crested weir has an error of less than 5% (95% confidence limit).

To convert the sill-referenced energy head into a water level head, an approach velocity coefficient is used: $C_v = (H_1/h_1)^u$ (Eq. 3.63). This C_v value was calculated for three control shapes with different values of the exponent u: rectangular u = 1.5, parabolic u = 2.0, and triangular u = 2.5. C_v was calculated as a function of the dimensionless ratio $\sqrt{\alpha_1}\, C_d\, A^*/A_1$, where A* is the area of the control cross-section below the water level in the approach channel. For these different control sections, the functions were found to be nearly identical. Consequently, for general water management practice, the same empirical relationships for C_d and C_v can be used for all shapes of the control section. The chapter concludes with an iteration procedure by which a programmable calculator can be used to produce the rating table of a long-throated flume or broad-crested weir.

Chapter 4 explains the concept of modularity and the related procedure of estimating the required head loss over a structure to maintain a unique relationship between the upstream sill-referenced head and the flow rate; i.e. the head loss required for the equations of Chapter 3 to remain valid. The approach followed is based on the assumption that this head loss can be divided into three parts: (i) the energy loss between the upstream head measurement section (gauging station) and the control section; (ii) the energy loss due to friction between the control section and the downstream head measurement section; and (iii) the losses due to turbulence in the diverging transition. Section 4.2 gives expressions for these separate parts of the required head loss and for the total head loss. Laboratory tests were used to verify the above theory. Recently published laboratory data on measured modular limits of flumes and weirs were compared with the values estimated by this procedure. The comparison shows that the procedure gives good results if the ratio $H_1/L \leqslant 0.5$. Photographs and water surface profiles are given as an aid in the visual detection of the modular limit (Section 4.5).

Chapter 5 discusses the demands imposed by irrigation and drainage practice and how these demands influence the design of a flume or weir. The structure may have three functions: (i) measurement of flow rate; (ii) controlled regulation of flow rate; and (iii) control of the upstream water level. Considering the period over which data need to be collected and the value of Q_{max}, it must be decided whether a temporary or permanent flume or weir should be used (Section 5.2). Sections 5.3 through 5.6 discuss the interrelation between the upstream freeboard requirement, the available head loss at the measuring site, the range of flow rates to be measured (Q_{max}/Q_{min}), the allowable error in the measurement of the flow rate, the reduction of losses in a downstream diverging transition, and the construction cost of the flume or weir. Section 5.6 also presents a relationship between the postulated value of Fr_1 at maximum flow and the area ratio A^*/A_1. This relationship is of special importance for a first tentative choice of a flume or weir that will be modular at the design maximum flow while meeting the postulated freeboard requirement. Section 5.7.3 describes exploratory laboratory tests on the influence of sediment on the head-discharge relationship of a flume and a weir. Both structures conveyed sediment equally well, but in the flume the head-discharge relationship was less disturbed than in the weir.

The final section of this thesis (Section 5.8) gives a 14-step iterative procedure for the design of long-throated flumes and broad-crested weirs. Based on the theory of Chapters 3 and 4, this procedure leads to a flume or weir that fits the selected measuring site and satisfies the demands of irrigation and drainage practice.

Samenvatting

Ten behoeve van een effektief waterbeheer zijn kunstwerken nodig waarmee het debiet in een grote verscheidenheid van kanalen en beken kan worden gemeten.
In Hoofdstuk 1 worden de meetgoten met lange keel en de lange overlaten geintroduceerd; hier wordt uitgelegd waarom deze familie van kunstwerken kan voldoen aan de eisen die voortvloeien uit de omgeving en hydraulica van de meeste meetlokaties.
Hoofdstuk 2 geeft een historisch overzicht van de meetgoten met lange keel en de lange overlaten. Het beschrijft hoe de hydraulische theorie van deze kunstwerken en hun vormgeving zich, afzonderlijk van elkaar, ontwikkelden. Het hoofdstuk besluit met een bespreking van recente pogingen tot het ontwikkelen van een theorie die algemeen geldig is voor elke meetgoot met lange keel of lange overlaat in combinatie met een willekeurige waterloop.

De overige hoofdstukken behandelen de onderdelen van een ontwerpprocedure (Paragraaf 5.8) die, voor elke willekeurige waterloop, leidt tot de hydraulische afmetingen en bijbehorende $Q-h_1$ relatie van een geschikte meetgoot of meetoverlaat. De belangrijkste onderdelen van deze procedure worden behandeld, namelijk de theorie van stromend water door kanaalvernauwingen van verschillende vorm en afmetingen (Hoofdstuk 3), de theorie en procedure benodigd om het verval, dat nodig is voor modulaire stroming, te voorspellen (Hoofdstuk 4), de randvoorwaarden die gerelateerd zijn aan de waterloop waarin het debiet moet worden gemeten, en de eisen die aan het kunstwerk worden gesteld met betrekking tot het traject waarover het debiet moet worden gemeten met een gevraagde nauwkeurigheid (Hoofdstuk 5).

Hoofdstuk 3 geeft twee fundamentele vergelijkingen:
$Q_i = A_c[2g(H_1-y_c)]^{0,50}$ (vlg. 3.12) en $y_c = H_1 - A_c/2\ B_c$ (vlg. 3.13), die algemeen geldig zijn voor kritische stroming van een ideale vloeistof door elke meetgoot met lange keel of over elke overlaat. Met gebruik van deze vergelijkingen kan de H_1-Q relatie worden afgeleid voor alle vernauwingen waarvoor eenduidige vergelijkingen bestaan voor A_c en B_c. Paragraaf 3.3 geeft zulke relaties voor enkelvoudig rechthoekige, driehoekige, trapeziumvormige, parabolische, cirkelvormige, en afgeknot-cirkelvormige vernauwingen en voor de samengestelde afgeknot-driehoekige en U-vormige vernauwingen. Voor andere vormen van een samengestelde vernauwing geeft Paragraaf 3.4 een procedure, waarmee deze H_1-Q relatie kan worden gevonden. Deze procedure is gebaseerd op de algemene eigenschap van deze familie meetgoten en overlaten, namelijk dat het ideale debiet
$Q_i = g^{0,50}\ A_c^{\ 1,50}/B_c^{\ 0,50}$ gelijk is voor elke vernauwing met een willekeurige vorm, mits waterspiegelbreedte B_c en de natte doorsnede A_c gelijk zijn. Het blijkt dat de H_1-Q_i relatie van een equivalente enkelvoudige vernauwing kan worden gebruikt voor bepaalde vernauwingen van samengestelde vorm.

Het ideale debiet wordt omgezet in een reëel debiet door gebruik te maken van een afvoercoefficient, $C_d = Q/Q_i$ (vlg. 3.59). Hierbij is aangenomen dat de waarde van C_d hoofdzakelijk wordt bepaald door wrijving langs en kromming van de stroomlijnen boven de keel van de meetgoot of de kruin van de overlaat (met lengte L). Een analyse van laboratoriumgegevens voor 105 meetgoten en overlaten van verschillende vormen en afmetingen en geconstrueerd van verschillende materialen, resulteerde in een grafiek voor de gemiddelde C_d waarde als een funktie van H_1/L. Deze grafiek kan worden toegepast onder de voorwaarde dat $0,1 \leqslant H_1/L \leqslant 1,0$, waarbij de fout in de waarde van C_d kleiner dan 5% is (95% betrouwbaarheidsinterval).
Door een toestroomsnelheidscoefficient $C_v\ (H_1/h_1)^u$ (vlg. 3.63) in te voeren kan, in plaats van de energiehoogte H, een overstorthoogte h_1

worden gebruikt voor het bepalen van de afvoerrelatie van een meetgoot of overlaat.
De waarde van C_v werd berekend voor drie vormen van de vernauwing waarvoor verschillende u-waarden gelden: rechthoekig: u = 1,5, parabolisch: u = 2,0 en driehoekig: u = 2,5. C_v werd berekend als funktie van de dimensieloze verhouding $\sqrt{\alpha_1}\, C_d\, A^*/A_1$, waarin A* gelijk is aan de oppervlakte van de vernauwing onder de waterspiegel in het toestromingskanaal. De C_v waarden voor deze drie vormen bleken nagenoeg gelijk te zijn. Ten gevolge hiervan mogen, voor algemene toepassingen bij waterbeheer, dezelfde relaties voor C_d en C_v worden gebruikt, ongeacht de vorm van de vernauwing. Dit hoofdstuk besluit met een iteratieve procedure, waarbij een programmeerbare zakrekenmachine kan worden gebruikt om de h_1-Q tabel te berekenen.

Hoofdstuk 4 behandelt het begrip modulair stromen en de daarbij behorende procedure om het verval over een kunstwerk te berekenen. Kennis van dit verval is nodig voor het vaststellen van het bestaan van een eenduidige relatie tussen overstorthoogte en debiet en daarmee voor de geldigheid van de vergelijkingen van Hoofdstuk 3. De hierbij gevolgde methode is gebaseerd op de veronderstelling dat dit benodigde verval kan worden gesplitst in drie delen: (1) het energieverlies tussen de doorsnede waar de overstorthoogte wordt gemeten en de doorsnede van de vernauwing waar het water kritisch gaat stromen, (2) het energieveries ten gevolge van wrijving tussen deze laatste doorsnede en de doorsnede waar de benedenstroomse waterstand (indien nodig) wordt gemeten en (3) het energieverlies ten gevolge van turbulentie in het vertragingsgebied.
Paragraaf 4.2 geeft vergelijkingen voor deze drie delen van het benodigde energieverval en voor het in totaal benodigde verval. Laboratorium- onderzoek werd verricht om deze methode te toetsen. Verder werden recent gepubliceerde laboratoriummetingen van de modulaire limiet van verschillende meetgoten en overlaten vergeleken met de volgens bovengenoemde procedure berekende limiet. Deze vergelijking bevestigde dat de procedure goede resultaten geeft, voor $H_1/L \leqslant 0{,}5$. Foto's en lengteprofielen van de waterspiegel zijn toegevoegd als hulpmiddel bij de visuele detectie van deze modulaire limiet (Paragraaf 4.5).

Hoofdstuk 5 bespreekt de eisen die de irrigatie- en drainagepraktijk aan een debietmeetkunstwerk stelt en hoe deze eisen het ontwerp van een meetgoot of overlaat beïnvloeden. Het kunstwerk kan drie funkties hebben: (1) meten van het debiet, (2) gecontroleerde beheersing van het debiet en (3) beheersing van het bovenstroomse peil. Gelet op de periode waarin debieten moeten worden gemeten en op de waarde van Q_{max}, moet worden beslist een tijdelijk dan wel een permanent meetkunstwerk te gebruiken (Paragraaf 5.2). De paragrafen 5.3 tot en met 5.6 bespreken de samenhang van de benodigde bovenstroomse waking, het beschikbare verval bij de meetlokatie, het meetbereik van de meetgoot of overlaat: Q_{max}/Q_{min}, de toelaatbare fout in de debietmeting, de beperking van het energieverlies in de vertragingszone en de bouwkosten van de meetgoot of overlaat. Verder geeft Paragraaf 5.6 het verband tussen de gewenste waarde van Fr_1 bij maximaal debiet en de oppervlakte verhouding A^*/A_1. Dit verband is een belangrijke schakel in de eerste keuze van een meetgoot of overlaat, die zowel modulair stroomt bij het maximale ontwerpdebiet als de vooropgelegde waking niet beperkt. Paragraaf 5.7.3 beschrijft een verkennend laboratoriumonderzoek naar de invloed van sediment transport op de h_1-Q relatie van een meetgoot en een overlaat. Beide kunstwerken verwerkten het aangevoerde sediment even goed, de h_1-Q relatie van de meetgoot werd echter minder verstoord dan die van de overlaat.

De laatste paragraaf (5.8) van dit proefschrift geeft een, uit 14 stappen bestaande, iteratieve procedure voor het ontwerpen van een meetgoot met lange keel of een lange overlaat. Gebaseerd op de in de Hoofdstukken 3 en 4 behandelde theorie, leidt deze procedure naar een meetgoot of overlaat die bij de gekozen meetlokatie past en voldoet aan de eisen uit de irrigatie- en drainagepraktijk.

Bibliography

Ackers, P., and Harrison, A.J.M. 1963. Critical depth flumes for flow measurements in open channels. Hydraulic Research Paper 5. 50 pp. Department of Industrial and Scientific Research, Hydraulic Research Station, Wallingford, U.K.

Arredi, F. 1963. Discussion of Adaptation of Venturi flumes to flow measurements in conduits' by H.K. Palmer and F.D. Bowlus. Proceedings of the American Society of Civil Engineers. Vol 62: 4: pp. 523-527.

Balloffet, A. 1951. Critical flow meters (Venturi flumes). Proceedings of the American Society of Civil Engineers. Vol 81: Paper 743.

Bazin, H.E. 1888. Expériences nouvelles sur l'écoulement en déversoir. Annales des Ponts et Chaussées. Mémoires et documents, 6 série, 16, 2 semestre, pp. 393-448, pl. 20-23. Paris.

Bazin, H.E. 1896. Expériences nouvelles sur l'écoulement en déversoir. Annales des Ponts et Chaussées. Vol. 7, pp. 249-357. Paris.

Bélanger, J.B. 1849. Notes sur le cours d'hydraulique. Mémoires Ecole Nat. des Points et Chaussées. pp. 32-33. Paris.

Blau, E. 1960. Die modelmässige Untersuchung von Venturikanälen verschiedener Grösse und Form. Veröffentlichungen der Forschungsanstalt fur Schiffahrt, Wasser und Grundbau 8. Akademie-Verlag, Berlin, German Democratic Republic.

Boiten, W. 1983. The trapezoidal-profiled broad-crested weir. S170-XI, April. 66 pp. Waterloopkundig Laboratorium, Delft.

Bos, M.G. 1974. The Romijn movable measuring/regulation weir. In: Small hydraulic structures. Irrigation and Drainage Paper 26: pp. 203-217. Food and Agriculture Organization, Rome.

Bos, M.G. (Ed.) 1978. Discharge measurement structures. 2nd ed. 464 pp. Publication 20, International Institute for Land Reclamation and Improvement (ILRI), Wageningen.

Bos, M.G. 1977a. The use of long-throated flumes to measure flows in irrigation and drainage canals. Agricultural Water Management, Vol 1: 2: pp. 111-126. Elsevier, Amsterdam.

Bos, M.G. 1977b. Discussion of Venturi flumes for circular channels' by M.H. Diskin. Journal of the Irrigation and Drainage Division, American Society of Civil Engineers. Vol 103: IR3: pp. 381-385.

Bos, M.G. 1979. Standards for irrigation efficiencies of ICID. Journal of the Irrigation and Drainage Division, American Society of Civil Engineers. Vol 105: IR1: pp. 37-43.

Bos, M.G. and Reinink, Y. 1981. Head loss over long-throated flumes. Journal of the Irrigation and Drainage Division, American Society of Civil Engineers. Vol.107: IR1: pp. 87-102.

Bos, M.G., Replogle, J.A., and Clemmens, A.J. 1984. Flow measuring flumes for open channel systems. 321 pp. John Wiley, New York.

Brakensiek, D.L., Osborn, H.B., and Rawls, W.J. (Coordinators) 1979. Field manual for research in agricultural hydrology. Agricultural Handbook 224. 550 pp. U.S. Government Printing Office, Washington, D.C.

British Standards Institution 1969. Methods of measurement of liquid flow in open channels. Part 4: Weirs and flumes; 4B: Long-base weirs. British Standard 3680. 39 pp. London.

British Standards Institution 1970. Methods of measurement of liquid flow in open channels. Part 4: Weirs and flumes; 4C: Flumes. British Standard 3680. 52 pp. London.

Chow, V.T. 1959. Open-channel hydraulics. 680 pp. McGraw-Hill, New York.

Clemmens, A.J., Bos, M.G., and Replogle, J.A. 1984. RBC broad-crested weirs for circular sewers and pipes. In: The Ven Te Chow Memorial Volume. (G.E. Stout and G.H. Davis, Eds.). Journal of Hydrology, Vol 68: 1: pp. 349-368.

Clemmens, A.J., Replogle, J.A., Bos, M.G. 1984. Rectangular flumes for lined and earthen channels. Journal of Irrigation and Drainage Engineering, American Society of Civil Engineers. Vol 110: 2: pp. 121-137.

Clemmens, A.J., Replogle, J.A., and Bos, M.G. 1984. FLUME : a computer model for estimating flow rates through long-throated measuring flumes. U.S. Dept. of Agriculture (in press).

Clemmens A.J., Bos, M.G., and Replogle, J.A. 1984. Portable RBC-flumes for furrows and earthen channels. Transactions of the American Society of Agricultural Engineers (in press).

Cone, V.M. 1916. A new irrigation weir. Journal of Agricultural Research, 13 March: pp. 1127-1143. U.S. Dept. of Agriculture, Washington, D.C.

Cone, V.M. 1917. The Venturi flume. Journal of Agricultural Research, Vol 4: 4: pp. 115-129. U.S. Dept. of Agriculture, Washington. D.C.

Cornell University 1899. Experiments of U.S. Deep Waterways Board. In: Weir experiments, coefficients and formulas. R.E. Horton, 1907: pp. 86-90. U.S. Government Printing Office, Washington, D.C.

Cornell University 1903, Experiments on model similar to Merrimac River Dam at Lawrence, Mass. In: Weir experiments, coefficients and formulas. R.E. Horton, 1907: pp. 107-109. U.S. Government Printing Office, Washington, D.C.

Delft Hydraulics Laboratory 1980. The V-shaped broad-crested weir. S170-VI, Jan. 105 pp. Waterloopkundig Laboratorium, Delft.

Diskin, M.H. 1976. Venturi flumes for circular channels. Journal of the Irrigation and Drainage Division, American Society of Civil Engineers. Vol. 102: IR3: pp. 383-387.

Dodge, R.A. 1982. Ramp flume model study : progress summary. 22 pp. U.S. Bureau of Reclamation, Denver.

Engel, F.V.A.E. 1934. The Venturi flume. The Engineer. Vol. 158, August 3: pp. 104-107; August 10: pp. 131-133.

Fane, A.B. 1927. Report on flume experiments on Shirhing Canal. Punjab Irrigation Branch, Paper 110: pp. 37-51, plate A-G. Punjab Engineering Congress, Bombay.

Food and Agriculture Organization 1979. Agriculture towards 2000 : a normative scenario. FAO, Rome.

Forchheimer, P. 1930. Hydraulik, Teubner, Leipzig and Berlin.

Gravelius, H. 1900. Herrn Basin's neue Untersuchungen über den Abfluss an Ueberfallen. Zeitschrift für Gewässerkunde, Vol. 3: 3: pp. 162-181. Leipzig.

Gulhati, N.D. 1955. Irrigation in the world : a global review. 130 pp. International Commission on Irrigation and Drainage, New Dehli.

Hall, G.W. 1962. Analytical determination of the discharge characteristics of broad-crested weirs using boundary layer theory. Proceedings of the Institution of Civil Engineers, London. Vol. 22: Paper 6607: pp. 177-190.

Harrison, A.J.M. 1967a. Boundary-layer displacement thickness on flat plates. Journal of the Hydraulics Division, American Society of Civil Engineers. Vol. 93: HY4: pp. 79-91; Closure: Vol. 95: HY3: pp. 1048-1051.

Harrison, A.J.M. 1967b. The streamlined broad-crested weir. Proceedings of the Institution of Civil Engineers, London. Vol. 38: pp. 657-678.

Harvey, W.B. 1912. Harvey's irrigation outlet. Punjab Irrigation Branch, Bombay.

Horton, R.E. 1907, Weir experiments, coefficients, and formulas. U.S. Dept. of the Interior, U.S. Geological Survey. pp.195. U.S. Government Printing Office, Washington, D.C.

Inglis, C.C. 1928. Notes on standing wave flumes and flume meter falls. 35 pp. Technical Paper 15. Public Works Department, Bombay.

International Standards Organization. 1980. Liquid flow measurement in open channels using flumes. Draft Standard ISO/DIS 4359. 53 pp. ISO, Geneva.

International Standards Organization. 1982. Liquid flow measurement in open channels : round-nose horizontal crest weirs. Standard 4374. 17 pp. ISO, Geneva.

International Standards Organization 1983. First draft proposal for an ISO standard on liquid flow measurement in open channels by weirs and flumes : trapezoidal profile weirs. By India. ISO/TC 113/SC 2N 329, 332, and 333. ISO, Geneva.

James, D.W., Hanks, R.J., and Jurinak, J.J. 1982. Modern irrigated soils. 235 pp. John Wiley, New York.

Jameson, A.H. 1925. The Venturi fLume and the effect of contractions in open channels. Transactions of the Institution of Water Engineers, London: 30 June: pp. 19-32.

Jameson, A.H. 1930. The development of the Venturi flume. Water and Water Engineering: March 20: pp. 105-107.

Kinghorn, F.C. 1975. Draft proposal for an International Standards Organization standard on the calculation of the uncertainty of a measurement of flow rate. Document ISO/TC 30/WG 14:24 E. ISO, Geneva.

Kraijenhoff van de Leur, D.A. 1972. Hydraulica I: 285-2C. 105 pp. Agricultural University, Wageningen.

Landbouwhogeschool. 1976. Aanvullend onderzoek V-vormige lange overlaat. Laboratorium voor Hydraulica en Afvoerhydrologie. Nota 35, April. 11 pp. Agricultural University, Wageningen.

Landbouwhogeschool. 1977. Debietmeting in een rioolbuis met behulp van een U-vormige vernauwing. Laboratorium voor Hydraulica en Afvoerhydrologie. Nota 41, October. 36 pp. Agricultural University, Wageningen.

Lindley, E.S. 1925. Canal falls and their use as meters. Punjab Engineering Congress, Lahore. Vol. 1: pp. 81-99.

Lindley, E.S. 1931. Measuring irrigation deliveries in the Punjab. Proceedings of the American Society of Civil Engineers. Vol. 37:2: pp. 269-283.

Mahbub, S.I., and Gulhati, N.D. 1951. Irrigation outlets. 184 pp. Atma Ram and Sons, Delhi.

De Marchi, G. 1937. Dispositivi per la misura della portata dei canali con minime perdite di quota. Parte III: Risultati delle esperienze. Vol. 14: pp. 189-215. l'Energia Elettrica, Milano.

Meijer-Peter, E., and Muller, R. 1948. Formulas for bed-load transport. Proceedings second meeting of the International Association for Hydraulic Research, Vol. 2, No. 2, pp. 39-64, Stockholm, Sweden.

Palmer, H.K., and Bowlus, F.D. 1936. Adaptations of Venturi flumes to flow measurements in conduits. Transactions of the American Society of Civil Engineers. Vol. 101: pp. 1195-1216.

Parshall, R.L. 1926. The improved Venturi flume. Transactions of the American Society of Civil Engineers. Vol. 89: pp. 841-851.

Peterka, A.J. 1958 (revised 1964). Hydraulic design of stilling basins and energy dissipators. 222 pp. U.S. Department of the Interior, Bureau of Reclamation, Washington. D.C.

Pitlo, R.H., and Smit, M. 1970. Discussion of Triangular broad-crested weir' by C.D. Smith and W.S. Liang. Journal of the Irrigation and Drainage Division, American Society of Civil Engineers.

Rafter, G.W. 1900. On the flow of water over dams. Proceedings of the American Society of Civil Engineers. Vol. 26: 3: pp. 226-319.

Replogle, J.A. 1970. Flow meters for water resource management. Water Resources Bulletin: Vol. 6: 3: pp. 345-374.

Replogle J.A. 1975. Critical flow flumes with complex cross-section. American Society of Civil Engineer. Specialty Conference Proceedings : Irrigation and Drainage in an Age of Competition for Resources. Logan, Utah, U.S.A. Aug. 13-15, pp. 366-388.

Replogle, J.A. 1977a. Discussion of 'Venturi flumes for circular channels' by M.H. Diskin. Journal of the Irrigation and Drainage Division, American Society of Civil Engineers. Vol. 103: IR3: pp. 385-387.

Replogle, J.A. 1977b. Compensating for construction errors in critical flow flumes and broad-crested weirs. In: Flow measurement in open channels and closed conduits. (L.K. Erwin, Ed.) National Bureau of Standards, Special Publication 434: Vol. 1: pp. 201-218. U.S. Government Printing Office, Washington, D.C.

Replogle, J.A. 1978. Flumes and broad-crested weirs : mathematical modeling and laboratory ratings. In: Flow measurement of fluids. (H.H. Dijstelberger and E.A. Spencer, Eds.): pp.321-328. North Holland Publishing Company, Amsterdam.

Replogle, J.A., and Bos, M.G. 1982. Flow measurement flumes: application to irrigation water management. In: Advances in Irrigation: Vol. 1: pp. 147-217. (D. Hillel, Ed.). Academic Press, New York.

Replogle, J.A., and Clemmens, A.J. 1979. Broad-crested weirs for portable flow metering. Transactions of the American Society of Civil Engineers. Vol. 22: 6: pp. 1324-1328.

Replogle, J.A., Clemmens, A.J., Tanis, S.W., and McDade, J.H. 1983. Performance of large measuring flumes in main canals. American Society of Civil Engineers. Specialty Conference Proceedings : Advances in Irrigation and Drainage : Surviving External Pressures. Jackson, Wyoming, July 20-22, pp. 530-537.

Robertson, A.I.G.S. 1966. Reprinted 1970. The magnitude of probable errors in water level determination at a gauging station. TN 7. Water Resources Board, Reading, U.K.

Robinson, A.R., and Chamberlain, A.R. 1960. Trapezoidal flumes for open-channel flow measurement. Transactions of the American Society of Agricultural Engineers. Vol. 3: 2: pp. 120-128.

Robinson, A.R. 1966. Water measurement in small irrigation channels using trapezoidal flumes. Transactions of the American Society of Agricultural Engineers. Vol. 9: 3: pp. 382-385, and 388.

Robinson, A.R. 1968. Trapezoidal flumes for measuring flow in irrigation channels. 15 pp. U.S. Dept. of Agriculture, ARS 41-141, U.S. Government Printing Office, Washington, D.C.

Romijn, D.G. 1932. Een regelbare meetoverlaat als tertiaire aftapsluis. De Waterstaatsingenieur, No. 9, Bandung.

Romijn, D.G. 1938. Meetsluizen ten behoeve van irrigatiewerken : handleiding door De Vereniging van Waterstaats Ingenieurs in Nederlandsch Indie'. 58 pp. Batavia.

Rouse, H., and Ince, S. 1957. History of hydraulics. 269 pp. Iowa Institute of Hydraulic Research, State University of Iowa, Ann Harbor.

Singer, J.C. 1936. Discussion of Adaptation of Venturi flumes to flow measurements' by H.K. Palmer and F.D. Bowlus. Transactions of the American Society of Civil Engineers. Vol. 101: pp. 1229-1231.

Skogerboe, G.V., Hyatt M.L., Anderson R.K. and Eggleston K.O., et al. 1967. Design and calibration of submerged open channel flow measurement structures. Part 3: Cut-throat flumes. Report WG 31-4, Utah Water Research Laboratory, Utah State University.

Smith, C.D., and Liang, W.S. 1969. Triangular broad-crested weir. Journal of the Irrigation and Drainage Division, American Society of Civil Engineers. Vol. 95: IR4: pp. 493-502; Closure: Vol. 97: IR4: 1971: pp. 637-640.

Smith, C.D. 1958. Open channel flow measurement with the broad-crested weir. Annual Bulletin, International Commission on Irrigation and Drainage. pp. 46-51. ICID, New Delhi.

Stevens, J.C. 1919. The accuracy of water-level recorders and indicators of the float type. Transactions of the American Society of Civil Engineers. Vol 83.

Thomas, C.W. 1957. Common errors in measurement of irrigation water. Journal of the Irrigation and Drainage Division, American Society of Civil Engineers. Vol. 83: IR2: pp. 1-24.

U.S. Bureau of Reclamation. 1957. Hydraulic and excavation tables. 11th edition. 350 pp. U.S. Dept. of the Interior. U.S. Government Printing Office, Washington, D.C.

Vierhout, M.M. 1973. On the boundary layer development in rounded broad-crested weirs with a rectangular control section. Laboratorium voor Hydraulica en Afvoerhydrologie, Rapport 3. 73 pp. Agricultural University, Wageningen.

Vlugter, H. 1941. De regelbare meetoverlaat. 8 pp. Waterloopkundig Laboratorium, Bandung.

Vries, M. de. 1973. Application of physical and mathematical models for river problems. 24 pp. Publication 112. Delft Hydraulics Laboratory.

Voortgezet onderzoek van registrerende waterstands meters, 1966. Hydraulica Laboratorium. 1966. Voortgezet onderzoek van registrerende waterstands meters. Nota 4. 15 pp. Agricultural University, Wageningen.

Wells, E.A., and Gotaas, H.B. 1956. Design of Venturi flumes in circular conduits. Transactions of the American Society of Civil Engineers. Vol. 123: pp. 749-771.

Woodburn, J.G. 1930. Tests of broad-crested weirs. Proceedings of the American Society of Civil Engineers. Vol. 56: 7: pp. 1583-1612. Also Transactions of the American Society of Civil Engineers. Vol. 96: 1932: pp. 387-408.

List of symbols

If subscripts are omitted, see list of subscripts at end of symbol list

A	=	cross-sectional area perpendicular to flow (flow area)
A^*	=	imaginary cross-sectional area at control section for water depth at same elevation as in approach canal
a	=	centrifugal acceleration of water particle
B	=	water surface or top width of flow
b	=	bottom width
C_d	=	discharge coefficient
C_v	=	approach velocity coefficient
D	=	average or hydraulic depth
D_a	=	characteristic particle diameter
d	=	canal depth
E	=	total energy head of particle with reference to an arbitrary elevation
EM	=	expansion ratio of diverging transition
F	=	canal freeboard
	=	centrifugal force
Fr	=	Froude number
f	=	focal distance for parabolic control shape
	=	friction coefficient
g	=	acceleration of gravity
H	=	energy head of flow referenced to weir sill
H_b	=	height of triangle in complex shape control
ΔH	=	loss in energy head over flume or weir
ΔH_L	=	energy loss due to friction over throat

ΔH_{trans} = energy loss due to friction over diverging transition

ΔH_{canal} = energy loss due to friction over part of tailwater channel

ΔH_f = energy loss due to friction over downstream part of structure

ΔH_d = energy loss due to turbulence in the diverging transition

h = head referenced to weir sill

Δh = change in sill referenced flow depth across flume

Δh_1 = difference between measured and true value of h_1

L = throat length

= length

L_a = distance from gauging station to start of converging transition

L_b = length of converging transition

L_d = length of diverging transition

L_e = length of tailwater channel from transition to fully developed flow (Section 2)

ML = modular limit

n = Manning roughness coefficient

P = pressure on water particle

p = sill height relative to channel bottom

Δp = change in sill height

Q = actual flow rate or discharge

Q_i = flow rate for an ideal fluid

ΔQ = change in flow rate

q = discharge per unit width

R = hydraulic radius (flow area/wetted perimeter)

r = radius of streamline curvature

s_f = hydraulic gradient

s_b = channel bottom slope

T = sediment transport capacity

u = exponent of head in the head versus discharge equation

= actual velocity of water particle

v = average velocity of flow

X_c = error in discharge from equations or rating tables

= error in discharge coefficient

X_{h_1} = error in upstreamsill referenced head

X_Q = errror in measured flow rate

Y = flow parameter for sediment transport

y = actual water depth

y_{sub} = flow depth at subcritical flow

y_{super} = flow depth at supercritical flow

Z = elevation of water particle

z = canal wall sideslope (horizontal to vertical)

α = velocity distribution coeffcient

γ = ratio of maximum to minimum flow rate to be measured with a structure

ϕ = angle made from center of pipe to edges of water surface in circular channel

ξ = energy loss coefficient for downstream transition

μ = ripple factor

π = pi = 3.1416....

ρ = mass density of water

ρ_r = relative density

ρ_s = mass density of sediments

ν = kinematic viscosity of fluid

θ = angle of opening for prismatic channels

Subscripts

1 = corresponds to head measurement section or gauging station

2 = corresponds to section in tailwater channel downstream from structure

c = corresponds to control section within weir or flume throat

min = corresponds to minimum design or anticipated flow rate

max = corresponds to maximum design or anticipated flow rate

Subject index

GPSR Compliance

The European Union's (EU) General Product Safety Regulation (GPSR) is a set of rules that requires consumer products to be safe and our obligations to ensure this.

If you have any concerns about our products, you can contact us on ProductSafety@springernature.com

In case Publisher is established outside the EU, the EU authorized representative is:

Springer Nature Customer Service Center GmbH
Europaplatz 3
69115 Heidelberg, Germany

Batch number: 10607320

Printed by Printforce, the Netherlands